Environmental Sociology

The third edition of John Hannigan's classic undergraduate text has been fully updated and revised to highlight contemporary trends and controversies within global environmental sociology. *Environmental Sociology* offers a distinctive, balanced treatment of environmental issues, reconciling Hannigan's much-cited model of the social construction of environmental problems and controversies with an environmental justice perspective that stresses inequality and toxic threats to local communities.

The author presents several key updates to the text, including:

- an all new introductory chapter, 'Planet in peril'
- a new case history chapter, 'Fear of fracking', focusing on shale gas exploration and drilling. This emergent energy/environmental issue helps students to gain a critical awareness of how risk construction, corporate power, politics and discourse creation interweave
- text boxes profiling environmental events, conflicts and activists in Canada, China, New Zealand, Southern and Western Africa and the United States
- recommended reading and suggested web links in each chapter.

The book concludes by examining the prospects for, and the nature of, a 'post-carbon society', making a strong case in favour of developing a new narrative of energy futures grounded in environmental economics, international relations and political ecology. This will be a key text for students of sociology, environmental studies and geography.

John Hannigan is Professor of Sociology at the University of Toronto and author of three major books: *Environmental Sociology* (Routledge, 2006); *Fantasy City: Pleasure and Profit in the Postmodern City* (Routledge, 1998) and *Disasters Without Borders: The International Politics of Natural Disasters* (Polity Press, 2012).

'John Hannigan's *Environmental Sociology* is a classic in environmental literature. It provides an invaluable lens into the social and cultural roots of the environmental issues of our day, challenging our worldviews about who we are as humans, how we relate to each other and how we understand the world of which we are a part. Hannigan's text captures the global urgency of the human impact on the environment and teaches us how environmental issues hold import for the way we structure our society. *Environmental Sociology* deserves a prominent place in any library that covers the environmental issues we care about.'

Andrew J. Hoffman, *Ross School of Business / School of Natural Resources & Environment, University of Michigan*

'The challenges of the century will come from the interaction of humankind with its environment. Hence the crucial importance of sociological reflections about the impact of society on the environment and the influence of the environment on human conduct. Hannigan's new edition of *Environmental Sociology* provides a distinctive, balanced and forward looking discussion of major environmental issues from a social science perspective.'

Nico Stehr, *Karl Mannheim Professor, Zeppelin University, Germany*

'Professor Hannigan again provides a fresh view of the social constructivist perspective as it challenges and extends understanding of society's interactions with nature and environmental problems. This new edition of *Environmental Sociology* reaches skilfully into uncharted territory in ways that will instruct and inspire students new to the field.'

Scott Frickel, *Associate Professor of Sociology, Brown University*

Environmental Sociology

Third edition

John Hannigan

Routledge
Taylor & Francis Group

LONDON AND NEW YORK

First published 1995
Second edition 2006
Third edition 2014
by Routledge
2 Park Square, Milton Park, Abingdon, Oxon OX14 4RN

and by Routledge
711 Third Avenue, New York, NY 10017

Routledge is an imprint of the Taylor & Francis Group, an informa business

British Library Cataloguing in Publication Data
A catalogue record for this book is available from the British Library

Library of Congress Cataloging in Publication Data
Hannigan, John A., 1948–.
  Environmental sociology / John Hannigan. – Third edition.
  pages cm
  Includes bibliographical references and index.
  1. Environmental sociology. I. Title.
  GE195.H36 2014
  363.7 – dc23
  2013037358

ISBN: 978-0-415-66188-1 (hbk)
ISBN: 978-0-415-66189-8 (pbk)
ISBN: 978-1-315-79692-5 (ebk)

Typeset in Perpetua
by Florence Production Ltd, Stoodleigh, Devon, UK

Printed and bound in the United States of America by
Edwards Brothers Malloy

To Ruth, as always

Contents

Illustrations

Figures

Tables

Preface and acknowledgements

In Lady Gaga's update on a classic Frank Sinatra song, she sings 'It's always harder/the second time around'. This is the third time around for *Environmental Sociology*, the first edition having been published in 1995. Writing this version has been the most challenging, but also the most rewarding of the three. Loyal and observant readers will notice that the front and back sections of the book look completely different. I've written a whole new introductory chapter, 'Planet in peril.' Environmental sociologists, I argue, have long been inclined to draw on an 'apocalyptic narrative', from William's Catton's concept of 'overshoot' to recent writing on 'runaway climate change'. In following this narrative thread, I offer a basic primer on some of the milestones of environmental thinking over the past 40 years – the Limits of Growth model from the 1970s, Jared Diamond's anthropological account of why societies 'collapse', John Urry's discussion of global climate change.

Several years ago, I attended a talk in the Geography Department at King's College, London by the noted Catalan economist Joan Martinez-Alier. At the reception following, one of the guests told me about 'fracking', that is, shale gas drilling, which she described as potentially devastating to the environment. Several days later, on the plane ride back to Canada, my wife insisted I watch a documentary film, 'Gasland'. Virtually unknown then, fracking has since grown to be a high-profile conflict, as evidenced by the recent 'Battle of Balcombe' in the south of England. In Chapter 9 of this book, 'Fear of fracking', I present a detailed case history of the shale gas controversy, employing the three-stage (assembling, presenting, contesting) model of environmental issue construction that has become the calling card of *Environmental Sociology*. We social constructionists are often accused (incorrectly) of claiming that environmental threats aren't 'real'. While scientists have yet to weigh in definitively, fracking looks to me like it has the capacity to unleash massive environmental damage, especially to the groundwater.

The third fresh chapter is the concluding one. The second edition of this book ended with my introducing an 'emergence model' of environment and society. I nudged this along further in a chapter of Redclift and Woodgate's (1997) second edition of the *International Handbook of Environmental Sociology*. I think this has great potential, but decided that this might be more fruitfully pursued in another venue. In the conclusion to this edition, I 'grasp the nettle', to use the title of a recent journal article I wrote, and address the question of whether and how environmental sociologists might best engage with the issue of 'low energy futures'. This is a topic that has provoked several extended discussions in sociology journals in recent years. I have come to think that energy futures are vitally important and should be the cornerstone of environmental sociology, as it enters its fourth decade. In particular, China's energy demands and policies will be crucial.

As was the case with the second, the impetus for this third edition was Gerhard Boomgaarden, Senior Publisher at Routledge. Gerhard has been especially mindful of making this book 'global' in scope as well as student friendly. This is especially reflected, I hope, in the addition of a number of new text boxes, most with an international dimension. Thanks also to his editorial assistant, Emily Briggs, who has kept the editorial process humming faultlessly throughout. Special thanks go to our son, Tim, for his technical assistance in preparing the final, merged manuscript. Most of all, thanks to my wife of 40 years, Ruth, for her love, encouragement and enthusiasm for this project.

Abbreviations

AEC	Atomic Energy Commission (US)
Alpac	Alberta-Pacific Forest Industries
AOU	American Ornithologists Union
AVIVE	Green Life Association of Amazonia (Brazil)
AYCC	Australian Youth Climate Coalition
BCP	Bioregional Conservation Planning
CBC	Canadian Broadcasting Corporation
CBD	Convention on Biological Diversity
CCB	Center for Conservation Biology (Stanford University)
CFCs	Chloroflurocarbons
CIPRs	Collective Intellectual Property Rights
CITES	Convention on International Trade in Endangered Species of Wild Fauna and Flora
CMS	Convention on Conservation of Migratory Species of Wild Animals
D.o.E.	Department of the Environment (UK)
DSP	Dominant Social Paradigm
EAAB	Empresa de Acueducto y Alcantarillado de Bogatá (Columbia)
EM	Ecological Modernization
EJM	Environmental Justice Movement
EJO	Environmental Justice Organization
ENGO	Environmental Non-Governmental Organization
EPA	Environmental Protection Agency (US)
EWG	Environmental Working Group (US)
FOEI	Friends of the Earth International
GREEN	Genetic Resource, Energy, Ecology and Nutrition Foundation (India)
HEP	Human Exemptionalism Paradigm
ICSU	International Council of Scientific Unions
IDNDR	International Decade for Natural Disaster Reduction
INBIQ	National Biodiversity Institute (Costa Rica)

IPCC	Intergovernmental Panel on Climate Change
IRI	International Research Institute for Climate Protection
IUCN	International Union for Conservation of Nature
LG	Limits of Growth
MAdGE	Mothers Against Genetic Engineering (New Zealand)
MDG	Millennium Development Goals
MDTP	Maloti-Drakensberg Transfrontier Conservation and Development Project (Lesotho/South Africa)
NAACP	National Association for the Advancement of Colored People (US)
NAS	National Academy of Sciences (US)
NASA	National Aeronautics and Space Administration (US)
NEP	New Ecological Paradigm
NGO	Non-Governmental Organization
NOAA	National Oceanic and Atmospheric Administration (US)
NSF	National Science Foundation (US)
NSMs	New Social Movements
ONAC	Office of Noise Abatement and Control
PCU	Project Coordination Unit
PVPA	US Plant Variety Protection Act
SARS	Severe Acute Respiratory Syndrome
SBSTA	Subsidiary Body for Scientific and Technological Advice
SCB	Society for Conservation Biology
SMO	Social Movement Organization
TFCA	Transfrontier Conservation Area
TKAG	Treasure Karoo Action Group (South Africa)
UCC	United Church of Christ
UFW	United Farm Workers
UNCED	United Nations Conference on Environment and Development
UNEP	United Nations Environment Programme
UNESCO	United Nations Economic, Social and Cultural Organization
UNFCCC	United Nations Framework Convention on Climate Change
USAID	US Agency for International Development
USDA	US Department of Agriculture
WMO	World Meteorological Organization
WRI	World Resources Institute
WWF	World Wildlife Fund

1 Planet in peril

When I was growing up in the 1950s and 1960s, the overriding threat to our planet was the nuclear arms race between the United States and the Soviet Union (Russia). Although we were not constantly paralyzed by fear, there were intermittent reminders that Armageddon constantly hovered on the horizon. During the inaugural Detroit (United States) – Windsor (Canada) International Freedom Festival in 1959, a popular attraction on the American side of the Detroit River was an exhibit of intercontinental ballistic missiles (minus nuclear warheads). Later on, as a teenager, I remember leaving the movie theatre feeling shaken after viewing *Dr. Strangelove or How to Stop Worrying and Love the Bomb* (1964), Stanley Kubrick's brilliant satirical film on thermonuclear war.

Two years earlier, during the Cuban Missile Crisis, the world had come the closest ever to an apocalyptic war. This was sparked by Soviet president Nikita Khrushchev's decision to place intermediate nuclear warheads on Cuban soil, threatening to deploy them if the United States invaded the island, as it had attempted the year before at the botched Bay of Pigs mission. Fortunately, after a tense week of negotiations, Khrushchev was convinced by US president John Kennedy to remove the missiles. Nevertheless, the spectre of nuclear warfare and total destruction continued to envelop the 1960s, only easing somewhat in 1968 when the Treaty on the Non-Proliferation of Nuclear Weapons was signed in Washington, Moscow and London.

As Boia (2005: 151–2) observes, during this era catastrophe scenarios, for the first time in history, went beyond cosmic and natural disasters such as those resulting from killer comets and super-volcanoes. It was instantly plausible, Boia says, to imagine catastrophic scenarios based on humanity's ability to trigger the forces of destruction. Furthermore, climatic upheavals and nuclear cataclysms go hand-in-hand. This is short-handed by the concept of the 'nuclear winter', wherein nuclear explosions create a dust cloud that covers the planet and blots out the sun's rays. Willis (2013) discusses a 'doomsday fixation' that blanketed this era, as symbolized by the 'Doomsday Clock' operated by the *Bulletin of Atomic Scientists* (Box 1.1)

Box 1.1 Doomsday Clock

After the scale of destruction wreaked by the dropping of atomic bombs in Hiroshima and Nagasaki, Japan became evident, many scientists involved in the Manhattan Project turned in horror away from the further development of nuclear weapons. Some of them founded the *Bulletin of Atomic Scientists*, a magazine dedicated to peace and security. In 1947 the *Bulletin* set up the 'Doomsday Clock' as a symbolic warning of the dangers of atomic weapons of war. The Clock was intended to evoke both the image of apocalypse (midnight) and the contemporary idea of nuclear explosion (countdown to zero).

Initially, the *Bulletin* set the clock at seven minutes to midnight, with midnight representing the end of the world. In 1953, after the test of the hydrogen bomb, the clock moved dangerously close to its endpoint, being adjusted at two minutes to midnight. 'Only a few more swings of the pendulum', *Bulletin* editors warned, 'and, from Moscow to Chicago, atomic explosions will strike midnight for Western civilization'. When the Cold War finally wound down in 1991, the setting was moved back to 17 minutes to midnight. As of August 2013, the hands are situated at five minutes to midnight. The editors explain, 'The challenges to rid the world of nuclear weapons, harness nuclear power, and meet the nearly inexorable climate disruptions from global warming are complex and interconnected' (*Bulletin of the Atomic Scientists*, September, 1953, cited in Willis 2013).

Limits of growth

As the 1970s unfolded, we were reminded the planet remained in a state of grave danger, but for an entirely different reason. This impending crisis is outlined in a seminal book *The Limits Of Growth*[1] (Meadows *et al.* 1972) compiled by four members (Dennis Meadows, Donella 'Dana' Meadows, Jorgen Randers, William Behrens) of Jay Forrester's systems dynamics group at the Sloan School of Management, Massachusetts Institute of Technology (MIT). The Club of Rome, a private international think tank founded in 1968 by Italian industrialist Aurelio Peccei commissioned the project and arranged funding through the Volkswagen Foundation. Harper (2008: 2010) claims that the Limits of Growth (LG) project, in combination with neo-Marxian political economy, constitutes 'the soil from which environmental sociology grew, and continues to inform its dominant modes of theorizing and empirical analysis'. Buell (2003: 144) attributes the tremendous importance of the book to the fact that it was the first to depict the global environmental threat as a 'social crisis' resulting from human economic and population growth and requiring nothing less than fundamental societal change.

Simply put, the LG authors hypothesized that the world's finite (only a fixed quantity is available) resources – timber, coal, oil – were being depleted at an alarmingly rapid clip and were in danger of running out. They pointed to runaway population growth, uncontrolled industrial production and material consumption as the chief culprits. The Meadows group deployed World3, a sophisticated computer simulation model of global population growth and economic development in order to generate seven different scenarios (possible futures). Patterns and trends of population growth, industrial output, resource availability and depletion, pollution and food production were all projected from 1970 levels all the way to the year 2100.

The most pessimistic scenario here anticipated nothing less than the total collapse of civilization and the advent of an exhausted and polluted planet by the mid or late twenty-first century. In the more optimistic scenarios, humans finally awake to the seriousness of the situation and take immediate and drastic action to bring the population and finite resources of the environment into a state of equilibrium. If this is not done, the Club of Rome report warned, then an *overshoot* of the limits of growth will occur followed by uncontrolled decline. The authors noted that technological solutions alone are not sufficient, as these can only serve to postpone the decline. Rather, technological improvements must be accompanied by changes that decrease the social, economic and political factors propelling growth.

What should be done about the approaching collision between population and material growth and the physical limits of the earth? Randers and Dana Meadows (1973) proposed a 'lasting solution', the cornerstone of which is a deliberate decision to stop population and capital growth. How this is to be achieved is never fully explained, although they ventured that, 'Perhaps only organized religion has the moral force to bring such a change, or perhaps it could come from an enlightened and widespread change in public education' (p. 351). Elsewhere, they seem to favour increasing the power of supranational bodies such as those associated with the United Nations.

To achieve a global equilibrium in accordance with the planet's physical limits, Randers and Meadows insist, people need to be freed from their preoccupation with material goods. Science and technology can play its role by 'developing ways of constructing products that last very long, do not emit pollution, and can be easily recycled'. Commercial competition need not be halted, but the total consumer market would be frozen and the 'emphasis would be on repair and maintenance rather than new production'. One positive offshoot of ending physical growth might be a more equitable distribution of wealth throughout the world.

Not everyone readily accepted this set of predictions, nor the solutions provided. As O'Riordan (1976: 60) explains, one important feature of the World3 computer model is that it is automatically programmed to catastrophe. That is, it is locked into a logic wherein growth in population, pollution, industrial pollution and the demand for food and raw materials are exponential, while the supply of the latter is finite. This is a 'recipe for doom' wherein the cataclysmic result is predetermined. There is no room here for unforeseeable advances in technology wherein the food and energy supplies can be significantly increased rather than remaining static. Human inventiveness and innovation are not factored into the model. One reason for this is ideological; the LG authors inherently distrust Western industrial experience and capitalist ideologies (Meadows *et al.* 1972, p. 63). The World3 model states unequivocally, 'Technological solutions designed to release some pressure caused by growth . . . can serve only to postpone the decline, if they are not accompanied by changes that decrease the social, economic, and political factors causing growth' (Meadows 1973: 43).

One of the first environmental sociologists to pick up on the importance of the LG model was William Catton, Jr. in his book *Overshoot: The Ecological Basis of Revolutionary Change* (1980). Catton bluntly advises his readers that 'even the "alarmists" who have been warning of the grave perils besetting [hu]mankind have not fathomed our present predicament' (1980: 5). He calls this a predicament rather than a crisis because the situation is neither temporary nor of recent origin. Of vital importance here is the concept of *carrying capacity*, defined as the maximum resource load beyond which the environment's ability to support life for a given kind of creature would be compromised. In the past, human beings have succeeded in taking over additional portions of the earth's total life-supporting capacity at the expense of other creatures. With the coming of an industrial society, technology became the primary, if temporary, means for augmenting human carrying capacity. Today, this band-aid solution is approaching its final days, as seemingly inexhaustible resources start to run out, even as the earth's population continues to grow. Employing an agricultural metaphor, Catton claims that humans are extending our carrying capacity by 'eating the seeds needed to grow next year's food'. Unless we act in a bold fashion, the inevitable result will be crash or die-off.

By the dawn of the new millennium, the crash foreseen by the MIT group and Catton seemed less probable.[2] This reflects the discovery of new energy sources and mineral resources, as well as significant increases in crop yields. Another possibility is that resource use and economic growth are becoming decoupled, that is, the economy continues to grow even as material consumption declines (Box 1.1).

In Vital Signs 2000, a widely consulted annual survey of environmental trends published by the independent, not-for-profit, Worldwatch Institute (WI), Lester Brown offered a gloomy but hopeful assessment of the earth's future. In 1999, the world population officially passed six billion, double the population in 1960. This demographic increase was most negatively felt in countries of the South. In country after country, Brown observes, the population was growing faster than the water supply. Furthermore, 'the demand for firewood and lumber was outrunning the sustainable yield of forests. And the demand for food was outrunning the cropland area' (Brown 2000: 17). Meanwhile, the world economy continued to expand.

The US$40.5 trillion worth of goods and services produced in 1999 was up more than six-fold from the US$6.3 trillion output of goods and services in 1950. This growth in the world economy threatens to outstrip 'the capacities of the earth's ecosystems to supply basic goods such as forest products, fresh water and seafood' (Brown 2000: 23). However, all was not devoid of hope. Citing a decline in world coal production and consumption accompanied by increases in natural gas and alternative energy sources, Brown notes:

> There were encouraging signs that the world is beginning to respond to the environmental threats that promise to undermine our future, but the gap between what we need to be doing to reverse the environmental deterioration of the planet and what we are actually doing continues to widen.
>
> (2000: 17)

Fast forward to the present. In *World on Edge: How to Prevent Environmental and Economic Collapse*, Brown (2011: ix) asks two key questions: if we continue with business as usual, how much time do we have left before our global civilization unravels, and how do we save civilization? Responding to the first question, he confesses that no one knows for sure, although it is very late in the day.[3] By way of a remedy, Brown proposes *Plan B*. This has four components: a massive cut in global carbon emissions of 80 per cent by 2020; the stabilization of world population at no more than eight billion by 2040; the eradication of poverty and the restoration of forests, soils, acquifers and fisheries (2011: 16).

Runaway climate change

In addition to food, energy, population and agricultural resource trends, *Vital Signs 2000* included a section on 'atmospheric trends'. Readers are alerted to a looming environmental crisis that is said to be endangering the planet – runaway climate change and global warming. In particular, the contributors claim that higher average temperatures since 1950, triggered by the unprecedented release of greenhouse gases by humans during the twentieth century, have resulted in more destructive storms and the melting of the earth's ice cover worldwide. As mountain glaciers shrink, large regions that rely on glacial runoff for drinking water – Lima, Peru is one notable example – face the threat of severe shortages (Masny 2000: 127).

In this view, over the past 150 years certain *greenhouse gases*,[4] notably carbon dioxide (CO_2), methane and water vapour, have become increasingly trapped in the earth's atmosphere. These gases act like a blanket, trapping heat in the infrared part of the spectrum and causing surface temperatures to increase on both land and the oceans. While greenhouse gases can emanate from natural sources such as rotting vegetation, it is steadily increasing industrial production that is said to be mostly to blame for the increase in CO_2. Warmer temperatures, in turn, provoke a long list of negative impacts: melting of the Arctic permafrost and mountain glaciers; catastrophic flooding in low-lying areas; severe drought and water shortages in the Southern hemisphere; significant biodiversity loss.

While scientific speculation into the greenhouse effect has been going on since the early nineteenth century, the topic first burst into public consciousness with the appearance on 23 June, 1988 before the US Senate Committee on Energy and Natural Resources of Dr. James Hansen. Hansen, a climatologist and director of NASA's Goddard Institute of Space Studies (GISS) in New York City, had initially been sensitized to the question of carbon warming while seeking to refine some preliminary calculations about high surface temperatures on the planet Venus compiled by Carl Sagan, an astronomer and astrophysicist who was widely known for co-hosting the vastly popular 1980 PBS television series *Cosmos*.

Hansen's testimony burst like a bombshell because he totally cast scientific caution aside. The earth, he declared, had entered a long-term warming trend and human-made greenhouse gases were almost surely responsible. 'It's time to stop waffling so much', Hansen asserted. 'It's time to say the earth is getting hotter.' This point was underscored by the record hot summer being experienced by Americans in 1988. Looking back on this event two decades later, Hansen said, 'Now as then, I can assert that these conclusions have a certainty exceeding 99 per cent' (Hansen 2008). In the aftermath of his testimony, Hansen became a media celebrity, making more than a dozen television appearances (Ungar 1992: 491).

In retrospect, Hansen (2008) observes, 'I was sure that time would bring the scientific community to a similar consensus'. Engineering this consensus fell to the Intergovernmental Panel on Climate Change (IPCC). First established in 1988 by the United Nations Environment Programme (UNEP) and the World Meteorological Organization (WMO) as an expert panel, the IPCC 'has had enormous impact upon world thinking about global warming' (Giddens 2009: 20). The IPCC Fourth Assessment Report (2007) proved to be especially dramatic. Whereas in its first three assessments, the human influence on global climate was judged to be 'discernible' or 'likely', the authors of the fourth report boldly stated, 'Most of the observed increase in global average temperatures since the mid-twentieth century is very likely due to the observed increase in anthropogenic greenhouse gas concentration'. According to IPCC conventions, 'very likely' implies a 90 per cent probability (Hulme 2009: 51).

Former US vice president Al Gore has been in the forefront in popularizing this grim forecast. Gore is featured in the Academy Award winning 2006 documentary *An Inconvenient Truth*. The film is built around a slick slide show lecture that Gore has been taking around the world in an effort to educate the public about the dangers of global warming. In one of the better-known clips from *An Inconvenient Truth*, Gore anticipates the collapse of a major ice sheet in Greenland or in West Antarctica, either of which could raise global sea levels by approximately 20 feet, flooding coastal areas and producing 100 million refugees. In *Our Choice: A Plan to Solve the Climate Crisis*, a written sequel to the documentary film, Gore claims that the stakes here are unprecedented: 'We have arrived at a moment unlike any other in history . . . What is at risk of being destroyed here is not the earth itself, of course, but the conditions that have made it hospitable for human beings' (Gore 2009: 16).

Bjorn Lomborg, whose book *The Skeptical Environmentalist* (2001) continues to be regarded as a 'declaration of war' (Pielke 2004a: 407) by many in the green movement, points to climate change as 'the environmental trump card'. He explains: 'Possibly we are not running out of raw materials, possibly we are actually doing better and better on almost any objective indicator, but if global warming demands a change, all other arguments will be of lesser impact' (2001: 258).

If the LG model was a dominant force in shaping environmental sociology during its formative years in the 1970s and 1980s, runaway climate change and global warming have been central to most conversations in the field since 1990. John Urry, a leading British sociologist, has written, 'Climate change, which may well be rapid and abrupt, constitutes a major transformation of human life and patterns of economic and social organization' (Urry 2010: 193). Global warming, Giddens (2009: 28) observes, is not the only danger created by human beings; other threats come from nuclear proliferation, Frankenstein nanotechnology, a food crisis or runaway population growth. Nevertheless, global warming stands out as an environmental danger as a consequence of its scale and potential future impact. York and Rosa (2012: 282) conclude that climate change, as well as biodiversity loss, nuclear waste and chemical toxins, 'represent severe threats to societal well-being due to their global scale and ubiquity'.

Frederick Buell (2003: 105–6) argues that global warming represents something radically different. Previously, it was thought that crisis problems, such as described in *The Limits of Growth* (Meadows *et al.* 1972), if left unchecked, might eventually lead to apocalypse, but this can be reversed. However, in the case of global warming, catastrophe could come suddenly and seemingly out of the blue. Having lived for several generations with environmental crisis, we now 'dwell in a world into which it is woven, intimately and everywhere' (p. 110). Like a small water leak that eventually brings down an entire dam, an alteration caused by changing climate could conceivably produce a cascade of changes that will bring down the whole system, its damage irreversible. Anguelovski and Roberts (2011: 19) conclude,

> Among all of the prevailing global environmental challenges, climate change is undoubtedly the most significant. It threatens our future development and, in some people's minds, puts at risk the continued existence of our own species and the global ecosystems on which we depend.

Collapse

One notable link between the earlier ecological models that spoke of eroding carrying capacity and overshoot and present-day catastrophic analyses of climate change is the best-selling book by Jared Diamond, *Collapse: How Societies Choose to Fail or Succeed* (2005). Writing in the journal *Current Sociology*, Constance Lever-Tracy (2008a: 457) recommends Diamond's book as 'surely an ideal starting point for a sociological debate and research programme about how an approaching ecological crisis could impact on society, and about the possibilities, likely agents and implications of alternative responses'.

Diamond, a biologically-trained geographer, conducts a postmortem on past societies that have 'collapsed' as a result of a catastrophic relationship with nature. A prime example is Easter Island. Located in the Pacific Ocean more than 3,200 km. west of Chile, it is best known for its more than 800 mysterious volcanic stone heads scattered over the island. For a long time no one could explain how the heads were transported and erected, until the Norwegian explorer and adventurer Thor Heyerdal (1958) demonstrated that timber was used to roll the statues and prop them up. By the sixteenth century, the island's inhabitants were impoverished. Even as the population increased, a limited and crucial resource, the palm forest, was decimated. In addition to their role in setting up the stone heads, wood from the forest was used for firewood and building materials.[5]

Another society that collapsed is Norse Greenland, the Greenland of the Vikings. Diamond observes that the environmentally-triggered collapse of Norse Greenland has parallels with the collapse of Easter Island. The Viking settlers squandered a host of fortunate circumstances that prevailed when they first arrived around 1000 BCE: a relatively mild climate, a virgin landscape that had never been logged or grazed, no immediate threat from aboriginal inhabitants (2005: 248). The newcomers burned woodlands to clear land for pasture, cut down the remaining trees for lumber and firewood, and prevented the regeneration of forested areas by permitting their livestock to run loose, grazing and trampling. They inadvertently made land even more useless by cutting turf for buildings and to burn as fuel. These actions damaged the environment by destroying the natural vegetation and by causing soil erosion. The coming of the Little Ice Age in the fifteenth century tipped a marginal agricultural economy over the edge by lowering hay production and clogging shipping lanes between Greenland and Norway with ice.

The parallels to the LG model here are readily evident. Increasing population in the face of limited resources leads to societal overshoot and crash. Diamond has been quite forthright in framing his case studies in the context of global warming. In the final chapter of *Collapse*, he recounts a familiar litany of disastrous outcomes associated with runaway climate change: a decrease in crop yields in already warm and dry areas, the disappearance of mountain snowpack decreasing the amount of

---

Box 1.2 'Peak Stuff'

In a recent research paper, independent researcher and author Chris Goodall (2011) analyzes figures prepared by the Office of National Statistics for the years 1993 to 2007 that represent estimates of the total inputs into the UK's material standard of living. These inputs are of three types: natural resources entering the economy (biomass, minerals, fossil fuels); key goods and services using these resources (food, paper and cardboard, textiles, fertilizers, cement, cars, energy use, travel and eventual waste (liquid and solid). Goodall found that Britain began to reduce its consumption of physical resources in the early years of the millennium, well before the economic slowdown that started in 2008. He argues that this reflects more than just the de-industrialization of Britain, a process that tends to cut resource use.

What accounts for this 'peak stuff' phenomenon, that is, the observation that beyond a certain level of economic development people simply stop consuming so much? Goodall acknowledges that he does not devote much attention to this question in his paper. He says that readers of the draft version produced their own hypotheses. In general terms, 'more sanguine commentators' hypothesize that technological innovation leads to greater efficiency and thus smaller inputs of fuels, minerals and biological materials (p. 2). More specifically, the UK economic growth figures may be skewed, insofar as the very highest income earners did much better than the rest of the population over the 15-year period. If so, there would not be a corresponding rise in the consumption statistics, insofar as the rich tend to invest their money, while the less privileged spend it on consumer goods. George Monbiot (2011), the environmental columnist for the *Guardian*, suggests that declining resource use may be the result of effective environmental campaigning, although he allows that this would be difficult to measure. Finally, it has been suggested that economic growth is decoupling from consumption because people are distancing their sense of wellbeing and satisfaction from consumerism. This 'anti-consumerism' hypothesis is equally difficult to validate, especially in the context of rising economic inequality in Britain.

---

water available for domestic use and irrigation, flooding and coastal erosion due to a rise in global sea levels (2005: 493–4). More generally, he suggests, 'human-caused climate change, the build-up of toxic chemicals in the environment and energy shortages will produce abrupt, potentially catastrophic effects in the 21st century' (Urry 2010: 208).

Consuming the planet to excess

Another common thread that links the limits of growth and runaway climate change narratives is the belief that consumerism, especially that practiced by the elite, centrally contributes to our planet's dire predicament and must be seen as integral to any solution. In its recent policy report, *Consumption and the Environment – 2012 Update*, the European Environment Agency points to the consumption of goods and services in EEA member countries as a major factor influencing global resource use:

> Although an increasing global population is a factor in rising pressures, it is consumption and production patterns in developed countries, with developing countries catching up rapidly, that are the key drivers of global environmental problems.
>
> (European Environment Agency 2012: 6)

Consumption, the report notes, creates both direct environmental pressures from the use of products and services, for example, through driving a car or heating a house, and indirect pressures that are created along production chains such as those related to food, clothing and the use of electricity. Most of these pressures are caused by private consumption in three sectors of the economy: housing, food and drink and mobility (especially that associated with tourism). A major reason why consumption negatively affects the environment and causes an over-use of resources is that many goods and services are too inexpensive. In America, fast food outlets offer large portions at a relatively low cost, while in Europe sun and sand winter holidays in the Costa del Sol and the Canary Islands are now more readily affordable.

In this spirit, the authors of the LG model concluded, 'we are inevitably faced with the necessity of recognizing that a larger population implies a lower material standard of living over the long term' (Meadows 1973: 44). Randers and Meadows (1973: 333) briefly touch on what the 'equilibrium society' of the future might look like in terms of consumerism: 'Freed from preoccupation with material goods, people might embrace the arts and other cultural activities and the enjoyment of unspoiled nature.' This new service society might not include fine dining, however, as the path to a global equilibrium may require that everyone be reduced to the minimum daily ration of calories, protein and vitamins while not permitting a minority to embrace a higher level of material consumption by enjoying 'fancy food'.

More recently, Urry (2010: 193) cites 'excessive' global consumption along with massive urban population growth, rising carbon emissions and multiple 'mobilities' (automobile use, long-distance air travel) as primarily responsible for destroying the *global* conditions of human life upon the earth. Urry is especially upset at the 'pervasive, mobilized, promiscuous commodification' represented by elite travel and tourism to leisure destinations such as Dubai. These places, he complains, 'are yet a further extension of the hyper-high-carbon societies of the 20th century, through their gigantic buildings, their profligate use of energy and water, and the vast use of oil to transport people in and out' (Urry 2010: 206).

Doomsday models such as that proposed by Urry and in *The Limits of Growth* assume that economic growth, the consumption of physical resources and strains on the global environment all go hand-in-hand. However, Goodall (2011) presents empirical evidence that seems to indicate that economic growth in a mature economy does not necessarily increase the pressure on the world's reserves of natural resources and on its physical environment (Box 1.2).

Constructing catastrophe

Nearly all of the writers and researchers that I have discussed thus far in this chapter (Lester Brown, William Catton, Jared Diamond, Anthony Giddens, Al Gore, James Hansen, Constance Lever-Tracy, Dennis Meadows, Eugene Rosa, John Urry, Richard York) fervently believe that the planet is in peril and that this is caused by some combination of capitalism, overpopulation, declining food production, dwindling natural resources, untrammeled industrialism, a failure to curb carbon emissions and toxic pollution, human greed and the pursuit of materialist lifestyles. Indeed, the language and logic of catastrophism has become part of the mainstream discourse of environmentalism. Last year, Carlos Bocuhy, who heads the Brazilian Institute of Environmental Protection, told an Associated Press reporter covering Rio+20, the 2012 UN Conference on Sustainable Development, 'We are facing the possibility of collapse if we don't change course' ('Pollution rife in Rio+20 host city 2012'). Accordingly, on 14 January, 2013 the *Bulletin of the Atomic Scientists* set the hands of the 'Doomsday Clock' at five minutes to midnight. In so doing, the *Bulletin* 'considered the current state of nuclear arsenals around the globe, the slow and costly recovery from events like Fukushima nuclear meltdown, and extreme weather events that fit in with a pattern of global warming' ('End near?' 2013).

Yet, 40 years after the publication of *Limits of Growth*, this view is far from a global given. Even though many scientists and environmentalists have identified a set of conditions leading to a perilous future, this hasn't guaranteed that their prognosis has been universally accepted or acted upon. Raymond Bryant, a British political ecologist, attributes the more or less 'business-as-usual' approach shown by most politicians and publics alike in a world of runaway climate change and apocalyptic capitalism to a 'lack of coherence and reasonableness in human thought and its ability to grasp an elusively alien world' (Bryant 2010: 179).[6]

Nadeau (2006) warns we have moved 'menacingly' close to a global environmental catastrophe. Yet, he complains, political leaders and economic planners continue to churlishly dismiss these apocalyptic scenarios as 'products of the overheated imaginations of muddleheaded idealists' (2006: 166). Furthermore, he attributes this to a failure to communicate this environmental crisis in terms that can readily be understood by those outside the scientific community. Nadeau's commentary suggests two separate explanations for our failure to recognize and act upon the dangers confronting our planet. First, he implies that the established economic and political orders deliberately obstruct change for their own selfish reasons. Alternatively, Nadeau raises the possibility that the prophets of ecological doom simply haven't done a very effective job in communicating and selling their message. In particular, he says, they have repeatedly failed to convince both political leaders and the public-at-large that scenarios of overshoot, collapse and runaway climate change are both real and pressing.

Catastrophe doesn't construct itself. Davis (2007: 8) notes that global warming was first recognized in 1896, but then dropped from sight until it re-emerged as an apocalyptic problem in the mid-1980s. Energy shortages were predicted in 1952 and 1956, then ignored until the 1973 'energy crisis' when OPEC (Organization of Petroleum Exporting Countries) quadrupled the price of oil and Middle Eastern oil producers embargoed shipments to the United States. This isn't to say that the pessimistic predictions of gurus such as Lester Brown and James Hansen can be summarily dismissed; however, convincing the world that catastrophe looms unless we act immediately and decisively is clearly no easy task.

In Chapter 3 of this book, I introduce a model of the 'social construction of environmental issues and problems'. According to this schema, there are three central tasks that characterize the process of defining environmental problems/ crises, bringing them to society's attention and provoking action.

First, environmental claims must be assembled, that is, they must be discovered, named and elaborated, something that most often occurs within the realm of science. As Jamison (1996: 224) notes,

> The contemporary concern with global environmental problems is due to scientific research: the hole in the ozone layer, the projections of global warming, and the implications of decreasing biodiversity have all been brought to light by scientists. Indeed, it seems fair to say that scientists have constructed these problems, and not just scientists, but particular cadres of well-supported and highly technified natural scientific researchers.

Furthermore, 'ownership' needs to be determined; otherwise, a problem will be free-floating with no one willing to acknowledge responsibility for initiating and promoting an action plan. The 40-member Club of Rome initially assumed this role in the 1970s with regards to the perceived threat of planetary collapse.[7] At Aurelio Peccei's urging, the Club invited MIT management professor Jay Forrester to apply his complex computer methodologies to the population-resources gap, funded the project and commissioned and promoted the *Limits of Growth*, as well as 30 more reports (Davis 2007: 31–6).

The tipping point with global warming occurred in October, 1985, at a technical workshop on greenhouse gases and climate variation convened in Villach, Austria and organized by the International Council of Scientific Unions (ICSU). What encouraged a breakthrough here was the decision to permit scientists to attend the workshop in an individual capacity rather than as formal representatives of their national governments. At Villach, the scientists collectively stated that 'the understanding of the greenhouse question is sufficiently developed that scientists and policy-makers should begin active collaboration to explore the effectiveness of alternative policies and adjustments' (WMO 1985: 3). This 'active collaboration' was evidently put in motion by Jim Bruce, second in command at the World Meteorological Association, who pressed hard for a political follow up to the collective statement issued at Villach. Within three years, a new body, the Intergovernmental Panel on Climate Change (IPCC), was up and running and climate change had been decisively propelled onto the international agenda (Paterson 1994: 175). Together with the impact of 'freak' weather conditions in Europe and America on public opinion, the consensus reached at Villach was instrumental in advancing the climate issue in science and bringing it onto the international political agenda (Skodvin 2000a: 146–7).

Second, it is crucial that environmental warnings be proclaimed beyond the cloistered world of scientists and policy analysts. Claims about the severity of a condition and what is needed to solve or mitigate the problem must be circulated and legitimated within the court of public opinion. Indeed, Jamison insists that something more than scientific construction is demanded here. Global environmental problems also require intermediaries such as NGOs (non-governmental organizations) to serve as information conduits between scientists, the media and the public, translate expert discourses into politics, and to recombine expert knowledge into policy-oriented packages (Jamison 1996: 224).

At this stage, scientific warnings are situated within a narrative or storyline. Storylines are 'powerful devices through which actors make sense of complex issues without recourse to comprehensive and cumbersome explanations' (Smith & Kern 2007: 5; cited in Davidson & Gismondi 2011: 26). Under the right set of conditions, 'big books' and other media texts (magazine articles, reviews, social media postings) can play a significant role in spreading new ideas from a critical community – a small group of critical thinkers who have developed an analysis of and solution to a perceived problem – into the broader culture (Meyer & Rohlinger 2012; Rochon 1998: 23). Nelissen *et al.* (1997) dedicate a section of their overview of classic texts in environmental studies to 'bestsellers' written in the first half of the 1970s.[8] These books were written for a large audience, published in paperback and promoted by the mass media. Most became 'a sort of bible for environmentalists' (Nelissen *et al.* 1997: 188). Furthermore, at a time when the first flush of environmental concern was fading, they were influential in convincing non-environmentalists of the importance of considering environmental problems from an ecological point of view.

Limits to Growth, the 1972 international bestseller, sold millions of copies in 29 languages and was a key element in exposing the general public to LG thinking. The original American publishers, Potomac Associates, presented a copy to every senator, member of the House of Representatives, governor and United Nations ambassador. An early copy of the manuscript was leaked to *Time* magazine, which immediately published a grim, doomsday article. Finally, the publisher organized a formal presentation and seminar at the Smithsonian Institution in Washington. Attendees included Chief Justice of the Supreme Court, Earl Warren; Philip Handler, president of the National Academy of Sciences; Wernher von Braun, a founder of the US space programme and Elliot Richardson, Secretary of Health, Education and Welfare (Atkisson 2011: 10).

Over the span of four decades, the project generated numerous spin-offs and updates; the most recent is *2052: A Global Forecast for the Next Forty Years* by Jorgen Randers (2012), one of the original *Limits* authors. Later on, Jared Diamond, who was already widely recognized after the success of his breakthrough bestseller *Guns, Germs and Steel*, situated concepts such as carrying capacity and societal collapse within a context that was easier for the public to fully grasp than the computer models employed by the *Limits* authors. The 'runaway climate change' threat has been proselytized by a number of high profile scientists (David King, James Hansen), politicians (Al Gore) and journalists (Elizabeth Kolbert, George Monbiot).

Contestation is the third stage in the social construction of environmental problems. Public recognition and media legitimacy are rarely enough to carry the day. While inspiring personal awareness and change is one important objective of environmental activism, successful resolution of a problem ultimately requires engaging in political campaigning, lobbying and making deals. This becomes especially challenging when the desired changes are economically, politically and socially controversial. Catastrophic claims such as those discussed in this chapter are especially difficult to translate into political action because they require an unprecedented degree of reallocation and structural change.

O'Riordan (1976: 65) observes that the Limits of Growth model may well have been a valuable catalyst to debate over growth and resource adequacy, but it fails 'to provide even a glimmer of how this formidable and fundamental revolution can actually be achieved'. He describes the authors' remedies as 'politically explosive'. These include 'a massive dose of non-growth' by means of a 40 per cent reduction in industrial investment, a 20 per cent disinvestment in agricultural activity and a 40 per cent fall in the birthrate. Today, in the midst of a prolonged worldwide economic crisis of unprecedented proportions, these remedies would seem even more difficult to justify politically. The same applies to proposed remedies to runaway climate change that include massive cuts in carbon use.

Environmental claims, especially those that call for a fundamental restructuring of the world order, are sharply contested. Despite becoming a bestseller, *The Limits of Growth* provoked a furious backlash in the early 1970s:

> The book came under fire from all sides. Scientists didn't like *Limits* because the authors, anxious to publicize their findings, put it out before it was peer reviewed. The political right rejected its warning about the dangers of growth.

The left rejected it for betraying the aspirations of workers. The Catholic church rejected its plea for birth control.

(MacKenzie 2012)

According to Alan Atkisson (2011: 13), a close friend of the Meadows, *Limits* was attacked by a 'small army' of prominent economists, scientists and political figures on the basis of its 'methodology, the computer, the conclusions, the rhetoric and the people behind the project'. Especially devastating reviews appeared in the *New York Times Book Review* and the journal *Science*.

More recently, the LG model has made somewhat of a comeback, in large part because of its relevance for theories of peak oil and of climate change. This has re-ignited old animosities, especially from conservatives. For example, Peter Foster, an editorial writer with the right-wing Canadian newspaper the *National Post*, recently wrote a column entitled '2052? More like 2084' in which he dismissed *The Limits of Growth* as 'pure junk science' that is 'based on the primitive zero-sum assumption that what was needed for the present poor and future generations to thrive was for the rich to abandon materialist lifestyles' (Foster 2012).

Much has been written about the politics of global warming (Giddens 2009; Hulme 2009). Environmental sociologists tend to treat this topic in a partisan and ideological manner (Dunlap *et al.* 2001). One reason is that both proponents and opponents tend to interpret climate change within the framework of pre-existing narratives concerning the moral trajectory of society (Rudiak-Gould 2012). Those who predict catastrophe due to uncontrolled greenhouse gas emissions visualize global warming within a narrative that casts contemporary society as locked into a downward spiral. By contrast, those who reject an apocalyptic vision of climate change subscribe to a narrative of progress wherein the world continues to be upward bound thanks to cumulative technological and material progress. These narratives make it even more difficult to negotiate the type of political compromises that are possible with environmental initiatives that are more limited in scope, for example, establishing a biosphere or conservation area.

Yearley (2009: 390–2) argues that there is a simple sense in which knowledge about climate change and the future climate is 'undeniably (and uncontroversially) socially constructed'. This knowledge involves an enormously complex system whose variables are not fully known or understood. In light of this imperfect knowledge, insight about the future depends primarily on the use of the method of 'climate modelling'. While technically sophisticated, climate modelling demands 'simplified versions of the atmosphere and the oceans'. Climate modellers rarely reside on the same page, nor are they perfectly in sync with other segments of the climate science community, notably geophysicists. This leads to conflicting scientific claims. Furthermore, climate projections (which have come to mean catastrophic projections) demand simplified versions of the behaviours of governments, corporations and consumers. These too are loaded with uncertainty, which is compounded by the fact that people may change their behaviours in response to the 'futures' predicted by the experts. Yearley (2009: 402) concludes that taking a constructionist approach to climate-change knowledge pays real dividends, especially insofar as it highlights the social functioning of the scientific community.

Climate change, Lee and Motzkau (2013: 462) observe, is best understood as 'an emergent biosocial phenomenon with the potential to diffract efforts to respond to it rather than as a well-defined problem'. By this they mean that every attempt to solve the global warming 'problem' only ends up mutating into something else: a problem of science communication and political controversy, a problem of international relations, a problem of energy sustainability, a problem of food security, a problem of runaway consumer consumption. Each of these carries with it a distinctive set of claims and social constructions.

In pondering the construction of climate change as a global problem, Beck and van Loon (2011: 114) point to a fundamental paradox. The dominant discourse on climate change, they argue, one that 'has a virtual monopoly in the sphere of world politics', points to a catastrophic *future* of rising sea levels, shifts in rain zones and rapidly advancing desertification. Yet, from a sociological perspective, the impact of climate change is already being felt here in the *present*. One major dimension of this is the shaping and reshaping of social and political landscapes, such that climate change politics become part of a *superpolitics* on a global scale. What emerges, they conclude (p. 113), is a discontinuity between the future of a catastrophic climate to come (science perspective) and the present anticipation of a catastrophic future (social science perspective).

Conclusion

I have chosen to begin our journey into environmental sociology by focusing on catastrophic scenarios, notably those associated with planetary overpopulation and collapse. These apocalyptic narratives stand out centrally in the field, from William Catton's foundational 1980 book *Overshoot*, to recent writing on global warming. Concern over environmental crisis is only one ingredient in the contemporary environmental movement. Within that movement, people have committed themselves to a wide variety of causes, such as wilderness appreciation and protection and the promotion of green lifestyles. Nevertheless, environmental optimism and concern about crisis stimulate each other, insofar as the prospect of a planet in peril intensifies 'people's impulses to experience, protect and cherish nature and work to ensure a viable future for human society' (Buell 2003: xi).

In the latter part of this chapter, I introduce my widely-cited model of the social construction of environmental issues and problems, which is elaborated in Chapter 3. According to this schema, there are three central tasks that characterize the process of defining environmental problems/crises, bringing them to society's attention and provoking action. Keeping in mind that 'catastrophe doesn't construct itself', I apply this constructionist framework specifically to the influential 'Limits of Growth' model from the 1970s, and to contemporary claims about 'runaway climate change'.

Notes

1   Dennis Meadows, Forrester's doctoral student, managed the research, while Meadows' wife Donella was the lead writer of *The Limits of Growth*, a simplified version of the technical report (Davis 2007: 32).

2   Nevertheless, the Meadows, in their 30-year update (Meadows *et al.* 2004) declared that humanity is still dangerously in a state of overshoot.

3   Answering his own question – how likely is it that our world will collapse? – the Dutch biologist Rob Hengeveld (2012: 292) concludes that this is 'theoretically inescapable', insomuch as 'system collapse follows from almost any simulation experiment', but he cannot specify when exactly this might occur.

4   Carl Sagan (1991: 212) notes that the phrase 'greenhouse effect' is actually a misnomer. Florists' greenhouses don't work through the operation of an 'infrared blanket', they work by preventing convective cooling. Nevertheless, the phrase is so widespread in atmospheric physics that we're stuck with it.

5   It is worth noting, however, that Easter Island was cursed by a dependence on the slow-growing Chilean Wine palm, which takes 40 to 60 years to mature, as compared to all the other Polynesian islands where fast-growing coconut and Fiji fan palms dominate (Lomborg 2001: 29).

6   Of course, this view doesn't allow for the possibility that the catastrophic scenarios dramatizing human–environment relations might be exaggerated or wrong.

7   In a report and proposal called 'The Predicament of Mankind', the Club of Rome identified a battery of interrelated global problems – uncontrolled population growth, disparity of wealth, social injustice, starvation and malnutrition, unemployment, crisis in democracy and civil unrest, decay of the city and depletion of natural resources – that would soon become uncontrollable if not fully understood and acted upon (Jain 2011: 59–60).

8   The books and articles included in this section of *Classics in Environmental Studies* are: Barry Commoner, *The Closing Circle* (1971), Donella Meadows *et al.*, *The Limits to Growth* (1972), Edward Goldsmith, 'A Blueprint for Survival' (1972), *Report of the United Nations Conference on the Human Environment* (1972), Kenneth Boulding, 'The Economics of the Coming Spaceship Earth' (1973), J. Tinbergen, *Reshaping the International Order* (1977), James Lovelock, *Gaia: A New Look at Life on Earth* (1979).

Further reading

Atkisson, A. (2011) *Believing Cassandra: How to be an Optimist in a Pessimist's World (2nd edition)*, London and Washington, DC: Earthscan.

Buell, F. (2004) *From Apocalypse to Way of Life: Environmental Crisis in the American Century*, New York: Routledge.

Diamond, J. (2011) *Collapse: How Societies Choose to Fail or Succeed (revised edition)*, New York, Penguin.

Urry, J. (2011) *Climate Change and Society*, Cambridge: Polity Press.

Online resources

Bulletin of Atomic Scientists. Available HTTP: www.thebulletin.org

Johnson, I. (2011) 'Are there limits to growth?' (video), Oxford Martin School. University of Oxford. Available HTTP: www.oxfordmartin.ox.ac.uk/videos/view/141

2 Environmental sociology

Key perspectives and controversies

Earth Day 1970 is often said to represent the debut of the modem environmental movement. Starting as a modest proposal for a national teach-in on the environment, it grew into a multi-faceted event with millions of participants. What most distinguished Earth Day, however, was its symbolic claim to be 'Day One' of the new environmentalism, an interpretation that was widely embraced by the American mass media, thus affording the environmental issue instant and widespread recognition (Gottlieb 1993:199).

When Earth Day inaugurated the 'Environmental Decade' of the 1970s, sociologists found themselves without any prior body of theory or research to guide them towards a distinctive understanding of the relationship between society and the environment. Indeed, according to Pretty *et al.* (2007: 2), they were somewhat taken by surprise by the environmental movement, and struggled to understand it; initially, the consensus was that 'an entirely new theoretical underpinning would need to be formulated'. While each of the three major classical sociological pioneers – Émile Durkheim, Karl Marx and Max Weber – arguably had an implicit environmental dimension to their work, this had never been brought to the fore, largely because their American translators and interpreters favoured social structural explanations over physical or environmental ones (Buttel 1986: 338). From time to time, isolated works related to natural resources and the environment appeared, mostly within the area of rural sociology, but these never coalesced into a cumulative body of work. In a similar fashion, social movement theorists gave short shrift to conservation groups, leaving historians to explore their roots and significance.

Why environmental sociology was slow to develop

To comprehend why environmental sociology was slow to develop, we must first understand how geographical and biological theories of social development and social change lost their predominance, even as sociology emerged as a distinctive discipline. There are two interrelated threads to this discussion: the rejection of geographical and biological determinism, and the rise to dominance of a distinctly *sociological* way of thinking that excluded non-social variables.

Backlash against geographical and biological determinism

In purging the 'determinist' perspectives that were popular among the early pioneers in the field, mainstream sociology also excluded nature and the environment. As Alexandrescu (2009: 57–8) notes, turn-of-the-century (and even later) sociological thinking was premised on the notion that a 'synthetic' science of society was possible. Early American sociologists of note such as William Graham Sumner and Franklin Giddings closely integrated their theories of social life with the well-established natural sciences (physics, geology, biology), as well as the other developing social sciences, notably economics and geography.

The impact of the geographical environment on the human condition was a topic of scholarly interest to these thinkers, who were inspired by the grand theories enunciated by the British historian Henry Thomas Buckle and the American geographer Ellsworth Huntington.

Buckle believed that the history of the world could be divided into two eras. Early on, the great Eastern civilizations – China, Egypt, India, Mesopotamia – flourished because they were blessed with warm temperatures and fertile soil. With a cheap and abundant food supply, population boomed, leading eventually to overcrowding, wide social disparities and the impoverishment of the masses. Needing to consume more food to survive in a colder climate, Europeans were necessarily more industrious and innovative. By modern times, Southern civilizations had stagnated, Buckle concluded, while those in the North had almost completely freed themselves from the burden of the environment through the power of science and technology (Boia 2005: 75).

No less modestly, Huntington attempted to prove that the rise and fall of entire civilizations such as that of ancient Rome follows the shift of the climatic zones in historical periods (Huntington 1917).[1] In so doing, Huntington proposed a theory of climate change and genetics, wherein he attempted to establish a series of correlations between climate and health, energy and mental processes such as intelligence, genius and willpower. Human progress was linked with a single ideal climate, that of temperate environments. As climate became steadily drier from the last ice age onwards, civilizations either migrated northwards or declined. In his own time, Huntington saw a 'productive climate' as one that encouraged industrial efficiency as measured by the number of pieces turned out by factory workers. Using this measure, Huntington strongly recommended that the United Nations headquarters be located in Providence, Rhode Island, which, according to his calculations, had the most productive climate in the world (Stehr & von Storch 2010: 51).

In assessing the worth of this 'geographical school', Sorokin (1964 [1928]: 192–3) refers to its fallacious theories, its fictitious correlations and its overestimation of the role of the geographical environment, but cautions that 'any analysis of social phenomena which does not take into consideration geographical factors is incomplete'. A fatal flaw here is the preoccupation with eugenics and racial stereotypes; indeed, Huntington was a leading member of the American eugenics movement throughout the 1920s and 1930s (Stehr & von Storch 2010: 31). Buckle, Huntington and their disciples fervently believed that the gentler environment of Western Europe and North America produced more 'civilized' cultures. By contrast, peoples of the tropics were more 'primitive'. Huntington, for example, noted that excessive humidity diminished energy, while 'dryness combined with heat induced a state of nervousness that went as far as loss of self-control' (Boia 2005: 100).

The natural world entered into early sociological discourse through the Darwinian concepts of evolution, natural selection, and survival of the fittest. In Darwin's theory, plants and animals are best suited to adapt to their environment and survive, while those less well equipped perish. The survivors pass on their advantages genetically to subsequent generations. Darwinism was seized upon by many of the early conservative sociological thinkers who applied its principles (not always accurately) to the human context (see Hofstadter 1959).

The most prominent social Darwinist was the English social philosopher Herbert Spencer, who proposed an evolutionary doctrine extending the principle of natural selection to the human realm. Spencer bitterly opposed any suggestion that society could be transformed through educational or social reform; rather, he believed that, if left alone, progress would evolve in a gradual fashion. Sumner was Spencer's greatest academic disciple in America, introducing his own concept of the 'competition of life' whereby humans struggle not just with other species for survival in the natural universe but also with each other in a social universe. Applying his theory to the *laisser-faire* capitalism of the day, Sumner legitimated the triumph of the 'robber barons', millionaire industrialists who made their money in banking, railroads and utilities through sharp and ruthless dealing. They were, he claimed, 'a product of natural selection', lions of business who would move society forward on the road to progress.

Both these 'single factor theories of social change' (Bierstedt 1981: 487) were rejected by mainstream sociology for largely the same reasons. By the 1920s, *laisser-faire* doctrines of the nineteenth century had given way to a new emphasis on social planning and social reform. 'Meliorism', the deliberate attempt to improve the well-being of members of society, flew in the face of these social theories that viewed social causation as unalterable, whether due to geography or biology. This was especially troublesome when it came to race.

Furthermore, the foundation of sociological theory had fundamentally shifted. Many sociologists came to accept psychology as the building blocks of sociology in place of physics or biology (Timasheff & Theodorson 1976: 188). This is evident in the social psychological tradition established by Mead, Cooley, Thomas and other

American symbolic interactionists who stress that the reality of a situation lies entirely in the definition attached to it by participating social actors. This definition in turn is socially shaped, as in Cooley's concept of the 'looking-glass-self' (1902). Physical (and environmental) properties become relevant only if they are perceived and defined as relevant by the actors (Dunlap & Catton 1992/3: 267).

By mid-century, the predominant theoretical outlook in American sociology had become structural functionalism, leaving behind in the dust the human ecological model formulated by Robert Park and his colleagues at the University of Chicago in the 1920s and 1930s (see the discussion later in this chapter). Functionalists carried forward Durkheim's idea that society constituted a social 'organism', which must constantly adapt to the outside social and physical environment. Its equilibrium or steady state could be knocked out of kilter by various disruptive events but it would always return to normal, just as the human body recovers from a fever. Alexandrescu (2009: 68) argues that functionalism prevented sociology from becoming concerned with the environment, insofar as it treated all social problems (including environmental ones) as nothing more than a temporary failure to adapt.[2] A clean environment was accorded little sociological value and treated as 'less than real' (ibid.).

Sociologists as hucksters for development and progress

Another explanation for sociological foot-dragging on environmental matters involves the world-view of sociologists themselves. In a steady stream of papers and articles from the late 1970s on, Catton and Dunlap argue that most sociologists during that era shared a fundamental image of human societies as being exempt from the ecological principles and constraints that govern other species. Sociologists favoured social engineering to achieve such goals as equality. At the same time, they accepted the possibility of endless growth and progress via continued scientific and technological development, while ignoring the potential constraints of environmental phenomena such as climate change (Dunlap & Catton 1992/3: 270). Mesmerized by the benefits of economic development and its sidekick, individual modernity, most sociologists at mid-century either completely ignored the natural environment or viewed it as something to be overcome with grit and ingenuity.

Some sociologists went even further, actively becoming advocates, even hucksters, for technological innovation and economic development. Nowhere was this more evident than in the literature on *modernization*, which was influential for two decades between 1955 and 1975. One leading illustration of this is Inkeles and Smith's book *Becoming Modern* (1974). For the authors, modernization denotes both a societal and personal transformation. Modern citizens, they argue, possess a repertoire of skills: they keep to fixed schedules, observe abstract rules, adopt multiple roles and empathize with others. Unlike those psychologically trapped in the past and unable to transcend traditional ways of thinking to become modern personalities, they are optimistic, opinionated, open to new experiences and consumers of information. These qualities are not inborn but must be acquired through life experience. The factory is the premier 'school of modernity'. It functions as a powerful model for rural migrants, inculcating, a sense of efficacy, a readiness for innovation and an openness to systematic change, respect for subordinates and the importance of planning and time. In this embrace of modernity, the influence of the physical environment is ignored. Being modern means advancing one's own personal goals rather than understanding and responding to natural forces. This view of nature is the antithesis of the 'ecological' ethic that stresses that human beings have no inherent claim to domination over nature but must coexist with other species on the earth.

That's not to say there weren't isolated critics of this pro-development paradigm, especially within the ranks of Marxist sociology; however, like religion, they tended to see the environment as a distraction from the necessity of class struggle. Even where the seriousness of environmental destruction was acknowledged, left-wing critics were inclined to focus on the class and power relations underlying this crisis rather than on factors relating more directly to the environment itself (Enzenberger 1979). Insomuch as Marxism eventually came to dominate social theory in some important regions of post-war European social theory, this resulted in the further exclusion of environmental issues from the discipline of sociology (Cotgrove 1991; Martell 1994).

Classical sociological theory and the environment

One possible source of inspiration for contemporary sociologists seeking to engage with environmental topics is the canon of classical social theory, notably that bequeathed to us by Durkheim, Weber and Marx. Each of these founders of the sociological field had something significant to say about nature and society, although this was often more implied than direct, and was embedded in the philosophical controversies and scholarly debates of the era in which they were writing.

Some commentators have been downbeat about the potential usefulness of this canon. Goldblatt (1996: 1–6) advises that we be wary of the legacy left to us by classical sociological theory insofar as it lacks an adequate conceptual framework with which to understand the complex interactions between societies and environments. As rewarding though it may be, Järvikoski (1996: 82–3) says, the reading of classic works authored by this triumvirate is simply not sufficient for adequate theorizing of contemporary environmental problems. Redclift (2000: 160) points out that the great names of classical social theory 'offer us no more than a few insights into the relatively new world of environmental policy-making'. Buttel (2000: 19) concludes that the legacy bequeathed by classical sociology is variable: some of the tools initially developed by the classical theorists are needed, but 'the overall thrust of the classical tradition was to downplay ecological questions and biophysical forces'.

On the other hand, there is a rich and expanding corpus of work in which environmental scholars seek to reveal this conventional wisdom to be premature. As we will see, some commentators (William Catton, John Bellamy Foster) deliberately adopt the strategy of extracting 'ecological' insights from the work of the classic thinkers that have been overlooked or misunderstood in the past. Others (Ray Murphy, Peter Dickens) are more inclined to smoke out concepts and ideas from the collected works of the sociological pioneers, even if these were not originally used in an environmental context, and apply them to the current environmental 'crisis' with some intriguing results. Some analysts have chosen to adopt a typological approach, organizing the field on the basis of classical theory. For example, Sunderlin (2003) defines and conceptualizes three key paradigms (individualist, managerial, class), each of which is derived from the classical sociological literature (Durkheim, Weber, Marx).

Émile Durkheim

Of the three founding figures in sociology, Durkheim is probably the least likely to be recognized as an environmental commentator.[3] In large part, this reflects his deliberate decision to elevate *social facts* over 'facts of a lower order' (i.e. psychological, biological).

For Durkheim, a social fact is 'any way of acting, whether fixed or not, capable of exerting over the individual an external constraint' (2002 [1895]: 117). This constraint is normally manifested in the form of law, morality, beliefs, customs and even fashions. We can verify the existence of a social fact, Durkheim ventured, by examining an experience that is characteristic. For example, children are compelled to adopt ways of seeing, thinking and acting that they otherwise would not have arrived at spontaneously.

Durkheim is quite firm in asserting that social phenomena cannot be explained through the lens of individual psychology. It is a central rule of the sociological method that 'the determining cause of a social fact must be sought among antecedent social facts and not among the states of individual consciousness' (p. 125). This rule may infuriate strong advocates of individualism, but no matter. Social facts, Durkheim insists, 'are consequently the proper field of sociology' (p. 112). While this vigorous defense of social facts and collective consciousness most certainly buttressed theoretical independence of sociology, it had the effect of warning off members of the new discipline from non-sociological approaches that were reductionist (i.e. they reduced explanation to biological or psychological factors) in nature. Nevertheless, Durkheim frequently utilized biological concepts and metaphors in the course of presenting his theory of societal transformation.

In *The Division of Labour in Society* (1893), he describes the evolution of modern societies from a state of *mechanical solidarity*, wherein social solidarity is a product of shared cultural values, to one of *organic solidarity*, where the social bond is a function of interdependence, most notably that arising out of an increasingly complex division of labour. Catton (2002: 92) proposes that Durkheim's theory was very much an attempt to devise a solution to what is essentially an ecological crisis of rising population paired with scarce resources (see Chapter 1). As societies became larger and denser, it would have been disastrous if everyone had continued to engage in agriculture. Increasingly, occupational specialization meant that the competition over arable land was lessened, even as that land became more productive thanks to technological innovation.

Durkheim was doubly hobbled, Catton says, both by his narrowly selective reading of Darwin and by the unavailability in the 1880s of our knowledge today of ecology and evolution (2002: 93). He erroneously supposed that Darwin believed increasing diversity to be a way of minimizing competition for scarce resources. Rather, Darwin had cautioned that co-evolution (two species evolving at the same time) could, in some cases, increase their *resemblance* to one another or result in one species bringing the other to extinction. In short, Darwin viewed specialization as a way in which one species could gain competitive advantage over another, not, as Durkheim believed, as a way of *lessening* rivalries and increasing mutual interdependence. Durkheim couldn't have been privy to the insights of modern ecology, which did not emerge as a sub-field of biology until the next century. Most crucially, no one in Durkheim's time recognized that mutual dependence was symbiotic but not necessarily balanced. That is, some interactions in nature benefit both member populations (*mutualism*) but others benefit one without either harming or benefiting the other (*commensalism*); and yet others are beneficial to one and detrimental to the other, as with predators and parasites (Catton 1992: 93). The latter gives rise to power differences, something especially significant when you are dealing with human ecological communities.

Max Weber

A second sociological pioneer whose work possesses an ecologically relevant component is Max Weber. Until recently, almost everything written about Weber and the environment came from two writers, Patrick West and Raymond Murphy. As Buttel (2002a) observes, West and Murphy locate this environmental connection in two entirely different corners of Weber's work.

In his doctoral thesis written in the 1970s, and in a later book chapter, West draws mostly on Weber's historical sociology of religion and his comparative research on ancient societies. He emphasizes that Weber analyzed concrete examples of struggles over natural resources, for example, the control of irrigation systems (West 1984). West's work remained obscure for more than a quarter of a century. Then, in 2012, John Bellamy Foster and Hannah Holleman published an article on Weber's approach to the environment in the *American Journal of Sociology* (Foster & Holleman 2012). The authors report that their analysis is influenced by West's dissertation, although they claim to approach the topic with more breadth and depth.

Foster and Holleman (2012: 1646) argue that Weber's sociology underscores the way in which industrial capitalism was enabled and constrained by the introduction of a process for smelting iron with coal. Ironically, this 'fateful union of iron and coal', as Weber called it, rescued the remaining forest land in Britain, insofar as the previously dominant method of charcoal-based iron smelting would have wiped out the woodlands in short order. The burning of fossilized fuel in blast furnaces and its use as a means to steam power constituted a major transformation for human society. However, Weber recognized that raw materials such as the mines and forests were not limitless. During a trip to the United States in 1904, Weber warned that the escalating scarcity of land and natural resources in that country would eventually constrict capitalism. Furthermore, he wrote about the pollution, filth, environmental degradation and wasted resources that he observed on his travels across America. In Chicago, for example, he noted that the pollution from the burning of soft coal was so severe that it was impossible to see further than three city blocks ahead (Foster & Holleman 2012: 1653).

Murphy's extended discussion of neo-Weberian environmental sociology is based largely on Weber's book *Economy and Society* (1978 [1922]). For Murphy, the key concept to be extracted here is *formal rationalization*. Rationalization is composed of several dynamic institutional components. Increased scientific and technical knowledge brings with it a fresh orientation in which nature exists only to be mastered and manipulated by humans. An expanding capitalist market economy leaves little room for anything beyond the calculating, self-interested pursuit of market domination. Industry and government are controlled by a bureaucratic apparatus, with a goal of attaining a high level of efficiency. The legal system operates like a technically rational machine. Together, these components promote a pervasive logic whereby efficiency reigns supreme, on occasion even superseding a sensible choice of goals or alternatives, what Weber called *substantive rationality*. Formal rationality thus dictates that the most efficient action is to clear-cut an old growth forest, even if this is in no way substantively rational from an ecological point of view (Murphy 1994: 29–30).

Murphy (1994: 34) identifies two interrelated processes first highlighted by Weber at the beginning of the twentieth century that have become distinctive features of our time: the *intensification of rationality* and the *magnification of rationality*. The more we try to run things according to the principle of dispassionate calculation the more we open the door to a swarm of unwanted and negative effects. When applied to the case of nature, this is called *ecological irrationality*. It is manifested in a wide range of destructive consequences from sensational technological disasters such as nuclear accidents to routine pollution events such as industrial dumping into urban storm sewers.

Drawing on another of Weber's (1946 [1918]) concepts – *intellectual rationality* – Freudenburg (2001) makes an important point about science, technology and risk. In contrast to tribal societies, the average individual in an industrial society cannot know more than a minimum about how technology works – unless he or she is a physicist, one who rides on the streetcar has no idea how the car happened to get into motion (Weber 1946 [1918]: 138–9). Consequently, while one may in principle master all things by intellectual calculation, in reality we depend on an army of experts to do so. Yet, as Freudenburg notes, this expectation is inherently problematic because a minority of the time these experts fumble the ball, leading to potential, and sometimes actual, environmental emergencies.

Karl Marx

Of the three main sociological traditions, it is that associated with Karl Marx that has provoked the most extensive response from present-day environmental interpreters. Marx and his early collaborator Friedrich Engels were only marginally concerned with environmental degradation per se but their analysis of social structure and social change has become the starting point for several formidable contemporary theories of the environment.

Marx and Engels believed that social conflict between the two principal classes in society, i.e. capitalists and the proletariat (workers), not only alienates ordinary people from their jobs but it also leads to their estrangement from nature itself. Nowhere is this more evident than in 'capitalist agriculture', which puts a quick profit from the land ahead of the welfare of both humans and the soil. As the Industrial Revolution proceeded through the eighteenth and nineteenth centuries, rural workers were removed from the land and driven into crowded, polluted cities, while the soil itself was drained of its vitality (Parsons 1977: 19). A single factor, capitalism, was held responsible for a wide range of social ills, from overpopulation and resource depletion to the alienation of people from the natural world with which they were once united. Marx and Engels saw the solution as an overthrow of the dominant system of production – capitalism – and the establishment in its place of a 'rational, humane, environmentally unalienated social order' (Lee 1980: 11).

Marx and Engels argue for the establishment of a new relationship between people and nature. However, it is not entirely clear what form such a relationship should take. In the work of the more mature Marx, this seems to follow a distinctly anthropocentric direction depicting humans as achieving mastery over nature, in no small part because of technological innovation and automation. This has been called a 'Promethean' (pro-technological, anti-ecological) attitude toward nature (Foster 1999: 372; Giddens 1981: 60).

By contrast, in Marx's early work the concept of the 'humanization of nature' is proposed. This suggests that humans will develop a new understanding of and empathy with nature. A key question here is whether this new understanding would be used solely for human emancipation or whether it would take a more ecocentric form in which the powers and capacities of non-humans would be enhanced. In the former case, the humanization of nature might be deployed to eliminate species and organisms that threaten human health (Dickens 1992: 86). As Martell (1994: 152) observes, the texts of the early Marx are too complicated and contradictory on ecological concerns to be the basis for a full-fledged theory of environmental protection; it may be more useful to pursue this project through other sources or frameworks.

Contemporary Marxist theory emphasizes not only the role of capitalists but also that of the state in fostering ecological destruction. Both elected politicians and bureaucratic administrators are depicted as being centrally committed to propping up the interests of capitalist investors and employers. While the incentive here is partly material, i.e. corporate campaign contributions, future job offers, public servants, politicians and capitalist producers are said to share an ethic that accentuates capitalist accumulation and economic growth as the dual engines which drive progress. This, they argue, holds for all political levels from the global system to the local community.

One widely noted reading of Marx's environmental views is John Bellamy Foster's seminal article on Marx's 'theory of metabolic rift'. According to Foster, Marx has been wrongly accused of providing minimal insight into the ecological crisis of our times. Indeed, due to the Promethean attitude that suffuses his later writing, he may even have impeded the understanding of environmental problems. To the contrary, Foster argues:

> Marx provided a powerful analysis of the main ecological crisis of his day – the problem of soil fertility within capitalist agriculture – as well as commenting on the other major ecological crises of his time (the loss of forests, the pollution of the cities, and the Malthusian specter of overpopulation). In doing so, he raised fundamental issues about the antagonism of town and country, the necessity of ecological sustainability, and what he called the 'metabolic' relation between human beings and nature.
>
> (1999: 373)

It is this latter point that Foster addresses most substantively. Borrowing from the vocabulary of mid-nineteenth-century chemistry, Marx employed the concept of *metabolism* to describe the complex interaction between society and nature. Metabolism, he observed, 'constitutes the fundamental basis on which life is sustained and growth and reproduction become possible' (Foster 1999: 383). By the 1860s, this organic relationship was being seriously undercut by the practices of capitalist agriculture. Most notably, landowners were accused of callously robbing the soil of its key nutrients by declining to recycle them. This, of course, is exactly what is still occurring, especially where monocultures (a single variety of a single crop grown for commercial profit) prevail. Marx describes this as a *metabolic rift* – the estrangement of human being from the natural world of the soil. This parallels the estrangement of workers from their labour and is attributable to the same source – capitalism.

Furthermore, Marx and Engels appear to have been early advocates of organic farming methods. For example, they write at length about the benefits of spreading manure on croplands, even suggesting that human waste from the city be recycled as fertilizer rather than polluting the rivers and oceans. Strangely enough, their inspiration for this view seems to have been the German agricultural chemist Justus von Liebig, who achieved renown as the inventor of synthetic fertilizers. By the late 1850s, Liebig had evidently come to the conclusion that soil depletion was becoming a major problem, especially in America where vast tracts of arable land were cultivated for the sole purpose of exporting grain to the big cities. Liebig even went so far as to recommend that the City of London organically recycle its sewage rather than dump it in the Thames river.

For Foster, the importance of Marx's theory of metabolic rift lies not just in his repatriation of Karl Marx as an advocate of organic agriculture but also in his successful application of sociological thinking to the ecological realm. Foster (1999: 400) calls this 'one of the great triumphs of classic sociological analysis' and proof that 'ecological analysis, devoid of sociological insight is incapable of dealing with the contemporary crisis of the earth'. Furthermore, it provides a portal through which contemporary environmental analysts might better understand the metabolic relation between humans and nature.[4]

Contemporary theoretical approaches to environmental sociology

Environmental sociology, Buttel (2003) observes, has gone through two distinct stages since its emergence in the 1970s as a discrete disciplinary area. In the first stage, the central task was to identify a key factor (or a closely related set of factors) that created an enduring 'crisis' of environmental degradation and destruction. More recently, there has been a significant shift towards another task: discovering the most effective mechanism of environmental reform or improvement, which will help 'chart the way forward to more socially secure and environmentally friendly arrangements' (p. 335).

In this section, I will begin by discussing two major approaches to the environment and society that were conceived with the first of these problematics in mind, and then proceed to an overview of two contrasting perspectives, reflexive modernization and ecological modernization, which address the second.

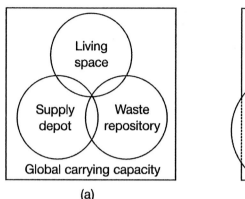

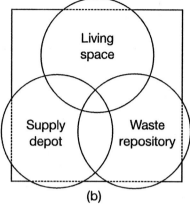

Figure 2.1 Competing functions of the environment: (a) *c*.1900; (b) current situation.
Source: Dunlap 1993.

Two foundational explanations for environmental degradation and destruction

In accounting for the causes of widespread environmental destruction, two approaches stand out: the ecological explanation as embodied in Catton and Dunlap's model of 'competing environmental functions', and the political economy explanation as found in Alan Schnaiberg's concepts of the 'societal-environmental dialectic' and the 'treadmill of production'. Both approaches view social structure and social change as being reciprocally related to the biophysical environment, but the nature of this relationship is depicted very differently (Buttel 1987: 471).

Ecological explanation

The ecological explanation for environmental destruction has its roots in the field of human ecology that remained dominant within urban sociology from the 1920s to the 1960s. Robert Park and his University of Chicago sociology colleagues introduced this human ecology model during the 1920s and 1930s. Park was well acquainted with the work of Darwin and his fellow naturalists, drawing on their insights into the interrelation and interdependence of plant and animal species. In his discussion of human ecology, Park [1936] (1952) begins with an explanation of the 'web of life', citing the once familiar nursery rhyme 'The House that Jack Built' as the logical prototype of long food chains, each link of which is dependent upon the others. Within the web of life, the active principle is the 'struggle for existence' in which the survivors find their 'niches' in the physical environment and in the division of labour among the different species.

If Park had been interested in the natural environment for its own sake, he might have realized that human intervention in the form of urban development and

Box 2.1 Harvesting Methane in Africa's 'Exploding Lake'

For a nation (Rwanda) chronically short of energy – only 1 in 14 homes have access to electricity – it seems a dream come true. Lake Kivu, on the border of the Democratic Republic of Congo and Rwanda, is the source of an estimated 55 billion cubic metres of methane gas, trapped along the lake bottom. Experts estimate that 90 per cent of the dissolved methane is 'harvestable' and could be worth as much as US$20 billion. This is enough to supply Rwanda's energy needs for 400 years, eliminating the need for burning wood for heat and cooking, a method that creates problems of air pollution and deforestation. Alas, extracting the methane brings with it some daunting risks and challenges.

At surface level, Lake Kivu is a tourist dream, with a beach and four-star resort framed by a mountain vista. It is a source of water, fish and sand for two million people. However, there are two linked problems. First, the 1,500-foot-deep lake contains large concentrations of methane and carbon dioxide. In recent years these gas levels have been increasing, probably from a combination of increasing volcanic activity and of organic sediment from the Ndakala, a sardine-like fish introduced nearly half a century ago. Second, Lake Kivu is situated in a region characterized by frequent seismic activity, within 15 miles of two active volcanoes. A geologic rift could pull apart and cause a crack. Large amounts of boiling lava entering the lake would lift the flexible lid provided by the lake waters, releasing clouds of deadly carbon dioxide and methane. In fact, this 'limnic eruption' actually occurred in the other two of Africa's 'killer lakes', Lake Monoun in Cameroon (1984) and nearby Lake Nyos (1986), where 1,700 people were suffocated by a cloud of carbon dioxide, rising from the lakes.

In 2008, the Rwandan government began work on a pilot project, whereby a pipe was dropped to the lake bottom, methane-rich, heavy water sucked up, the gas separated out, purified and then piped to a new power plant. With some encouraging results in hand, a number of private energy companies are contemplating larger projects, most notably Contour Global, a New York-based company that received the 2011 Africa Power Deal of the Year award from *Euromoney's Project Finance* magazine.

Sources: Browne (2010); Cowan (2008); Halper (2012); IRIN (2007); Rice (2010)

industrial pollution artificially break this chain, thereby upsetting the 'biotic balance'. In fact, he acknowledged that commerce, in 'progressively destroying the isolation upon which the ancient order of nature rested', has intensified the struggle for existence over an ever-widening area of the habitable world. However, he believed that such changes had the capacity to give a new and often superior direction to the future course of events forcing adaptation, change and a new equilibrium.

Biological ecology was principally a source from which Park borrowed a series of principles, which he then applied to human populations and communities. In doing so, he notes that human ecology differs in several important respects from plant and animal ecology. First, humans are not so immediately dependent upon the physical environment, having been emancipated by the division of labour. Second, technology has allowed humans to remake their habitat and their world rather than to be constrained by it. Third, the structure of human communities is more than just the product of biologically determined factors; it is governed by cultural factors, notably an institutional structure rooted in custom and tradition. Human society, in contrast to the rest of nature, is organized on two levels: the biotic and the cultural.

Park and his colleagues applied their principles of human ecology to the processes that create and reinforce urban spatial arrangements. They saw the city as the product of three such processes: (1) concentration and de-concentration, (2) ecological specialization and (3) invasion and succession. The building blocks of the city were said to be 'natural areas' (slums, ghettoes, bohemias), the habitats of natural groups that were in accordance with these ecological processes. The city was depicted as a territorially-based ecological system in which a constant Darwinian struggle over land use produced a continuous flux and redistribution of the urban population. Nowhere was this more evident than in the *zone in transition*, an area adjacent to the central business district, which went from a coveted residential district to a blighted area characterized by low rent tenants, deviant activities and marginal businesses.

An important issue here is whether the notion of an ecosystem should be accepted at face value or merely treated as an analogy. It seems likely that Park and the Chicago School had the latter in mind, adopting the conceptual language of biological ecology because it was the scientific flavour of the day. Alexandrescu (2009: 64) points out that that despite their guiding dictum – 'society functions like nature' – Chicago human ecology portrays it in coldly economic terms as 'the realm of blind and impersonal human experience'. Others took the ecological metaphor more literally. For example, the noted economist Kenneth Boulding (1950: 6) claimed that he was using the ecosystem concept in its proper sense, and not merely as an analogy. Society is, he wrote, 'something like a great pond' filled with 'innumerable "species" of social life, organizations, households, businesses and commodities of all kinds' (1950: 6).[5]

Spaaragen *et al.* (2000b: 4) have observed that the human ecology tradition stood at the birth of environmental sociology. Specifically, they cite Riley Dunlap and William Catton's *new ecological paradigm* (NEP) as 'one great effort to redefine the relationships between human societies and their natural environments'.

In my own estimation, the ecological basis of environmental destruction is probably best described in Catton and Dunlap's *three competing functions of the environment* (see Figure 2.1). This scheme has been much less widely disseminated than their NEP theory even though it is more conceptually interesting.

Catton and Dunlap's model specifies three general functions that the environment serves for human beings: supply depot, living space and waste repository. Used as a supply depot, the environment is a source of renewable and non-renewable natural resources (air, water, forests, fossil fuels) that are essential for living. Overuse of these resources results in shortages or scarcities. Living space or habitat provides housing, transportation systems and other essentials of daily life. Overuse of this function results in overcrowding, congestion and the destruction of habitats for other species. With the waste repository function, the environment serves as a 'sink' for garbage (rubbish), sewage, industrial pollution and other byproducts. Exceeding the ability of ecosystems to absorb wastes results in health problems from toxic wastes and in ecosystem disruption.

Additionally, each of these functions competes for space, often impinging upon the others. For example, placing a garbage (rubbish) landfill in a rural location near to a city both makes that site unsuitable as a living space and destroys the ability of the land to function as a supply depot for food. Similarly, urban sprawl reduces the amount of arable land that can be put into production while intensive logging threatens the living space of native (aboriginal) peoples. As seen in Box 2.1, extracting dissolved methane from Lake Kivu, Rwanda in order to generate cheap electricity could cripple fish stocks, destroy the local economy and potentially trigger a volcanic catastrophe. In recent years, the overlap, and therefore conflict, among these three competing functions of the environment has grown considerably. Newer problems such as global warming may be attributed to competition among all three functions simultaneously.

There are several attractive advantages to Catton and Dunlap's competing functions of the environment. First, it extends human ecology beyond an exclusive concern with living space – the central focus of urban ecology – to the environmentally relevant functions of supply and waste disposal. In addition, it incorporates a time dimension: both the absolute size and the area of overlap of these functions are said to have increased since the year 1900.

But there are problems with the model. There is no evidence of a human hand here. It says nothing about the social actions involved in these functions and how they are implicated in the overuse and abuse of environmental resources. Significantly, there is no provision for changing either values or power relationships. This is puzzling, since one would have thought that Catton and Dunlap would have attempted to link their ecological model to the new human ecology as emphasized in the HEP/NEP contrast. Freudenburg and Wilkinson (2008: 8) call for the Dunlap-Catton typology to be expanded to add a fourth function: the use of the biophysical environment as a location and set of channels for distributing social inequalities. The problem today, they say, isn't just that a need for waste disposal sites is bumping up against a need for living spaces; rather, some people are disproportionately benefitting from occupying prime properties in affluent neighbourhoods, while disposing of their wastes in or near the living spaces of marginalized folk.

Finally, one cannot help comparing the longitudinal features of the Catton-Dunlap model to Beck's (1992) explanation of the transformation from an industrial to a risk society (see the next section of this chapter). Both models recognize some of the same features: the increasing globalization of environmental dangers, the rising prominence of output or waste-related elements as opposed to input or production-related ones. However, Beck's model is ultimately more exciting because it centrally incorporates the process of social definition. Beck's (1992: 24) criticism of environmental risk assessment, i.e. that 'it runs the risk of atrophying into a discussion of nature without people, without asking about matters of social and cultural significance', is equally applicable to Catton and Dunlap's competing functions of the environment.

Political economy explanation: the 'societal-environmental dialectic' and the 'treadmill of production'

Within environmental sociology, the most influential explanation of the relationship between capitalism, the state and environment can be found in Alan Schnaiberg's book, *The Environment: From Surplus to Scarcity* (1980). Drawing on strands of both Marxist political economy and neo-Weberian sociology, Schnaiberg outlines the nature and genesis of the contradictory relations between economic expansion and environmental disruption.

Schnaiberg depicts the political economy of environmental problems and policies as being organized within the structure of modern industrial society, which he labels the *treadmill of production*. This refers to the inherent need of an economic system to continually yield a profit by creating consumer demand for new products, even where this means expanding the ecosystem to the point where it exceeds its physical limits to growth or its carrying capacity. One important tool in fuelling this demand is advertising, which convinces people to buy new products as much for reasons of lifestyle enhancement as for practical considerations.

Schnaiberg depicts the treadmill of production as a complex self-reinforcing mechanism whereby politicians respond to the environmental fall-out created by capital-intensive economic growth by mandating policies that encourage yet further expansion. For example, resource shortages are handled not by reducing consumption or adopting a more modest lifestyle but by opening up new areas to exploitation.

Schnaiberg detects a *dialectic tension* that arises in advanced industrial societies as a consequence of the conflict between the treadmill of production and demands for environmental protection. He describes this as a clash between 'use values'; for example, the value of preserving existing unique species of plants and animals, and 'exchange values', which characterize the industrial deployment of natural resources.

As environmental protection has emerged as a significant item on the policy agendas of governments, the state must increasingly balance its dual role as a facilitator of capital accumulation and economic growth and its role as environmental regulator and champion. On occasion, governments find it necessary to engage in a limited degree of environmental intervention in order to stop natural resources from being exploited with abandon and to enhance its legitimacy with the public. For example, in the progressive era of American politics in the late nineteenth and early twentieth centuries, the US government responded to uncontrolled logging, mining and hunting on wilderness lands by expanding its jurisdiction over the environment. Especially under the presidency of Theodore ('Teddy') Roosevelt, it created national forests, parks and wildlife sanctuaries, set limits and rules for the use of public lands and restricted the hunting of endangered species. It did so, however, as much out of a desire to increase industrial efficiency (Hays 1959), regulate competition and ensure a steady supply of resources (Modavi 1991) as it did from any sense of moral outrage. Similarly, the sudden emergence of toxic waste as a premier media issue in the early 1980s led to Congressional efforts in the United States to pass a new 'Superfund' law that would give the government statutory authority and the fiscal mechanisms to undertake clean-up operations without first having to legally identify the responsible parties. This was, Szasz (1994: 65) notes, not simply a matter of lawmakers addressing a newly recognized social need, but instead 'one of those quintessential "time to make a new law" moments so characteristic of the American legislative process'.

Nevertheless, most governments remain wary of running the risk of slowing down the drive towards economic expansion or decelerating the treadmill of production (Novek & Kampen 1992). Caught in a contradictory position as both promoter of economic development and as environmental regulator, governments often engage in a process of *environmental managerialism* (Redclift 1986), in which they attempt to legislate a limited degree of protection sufficient to deflect criticism but not enough to derail the engine of growth. By enacting environmental policies and procedures that are complex, ambiguous and open to exploitation by the forces of capital production and accumulation (Modavi 1991: 270) the state reaffirms its commitment to strategies for promoting economic development.

Other more stridently left-wing critiques have been even more unsparing in linking the dynamics of capitalist development to the rise in environmental destruction. Geographer David Harvey (1974) accuses capitalist supremos of deliberately creating resource scarcities in order that prices may be kept high. Faber and O'Connor (1993) charge that the goal of capital restructuring in the 1980s and 1990s, which includes geographical relocation, plant closures and down-sizing, was to increase the exploitation of both the workers and nature; for example, by reducing spending on pollution control equipment. Cable and Cable refuse to rule out the possibility of insurrection in the United States if the grievances of grassroots environmental groups continue to be ignored by capitalist economic institutions (1995: 121). Schnaiberg himself (2002: 33) complains that the central tenets of the treadmill have not found their way into the environmental sociological literature in any significant way because they are too 'radical'. That is, if the treadmill is indeed operating as he describes, then it can only be slowed down by a major and sustained political mobilization akin to a revolution, something that would be sharply resisted by politicians, government agencies and corporate America.

Subsequently, the 'treadmill of production group'[6] has addressed the application of the treadmill of production to a global context. Ignoring the negative environmental impacts that the treadmill has produced in less developed regions, the leaders of Southern nations, in concert with the governments and corporations of the North, have sought to reproduce industrialization as experienced by the First World. The primary mechanism for achieving this is the transfer of modern Western industrial techniques from North to South (Schnaiberg & Gould 1994: 167). As Redclift (1984) and others have noted, this transplant has been largely unsuccessful both in economic and environmental terms. Dependency on global markets has made economic development a risky venture for many Southern nations, especially where these markets can easily be decimated by the appearance of new, low-cost alternatives elsewhere in the world. Furthermore, development schemes require an expensive infrastructure of roads, hydroelectric power dams, airports, etc., which must be paid for by borrowing heavily from northern financial institutions. Such projects often fail to produce the expected level of economic growth while at the same time causing massive ecological damage in the form of flooding, rainforest destruction, soil erosion and pollution.

The treadmill of production explanation has the advantage of locating present environmental problems in the inequities of humanly constructed political and economic systems rather than the abstract conflict of functions preferred by human ecologists. This brings it closer to the orbit of mainstream sociological theory than does the more idiosyncratic approach advocated by Catton and Dunlap. The concept of the treadmill is unique insofar as it is based in sociological reasoning, but, at the same time, features a key or penultimate dependent variable – environmental destruction – that is biophysical. This makes it 'the single most important sociological concept and theory to have emerged within North American environmental sociology' (Buttel 2004: 323).

Schnaiberg complained that his treadmill of production model did not achieve the paradigmatic status within environmental sociology that he would have liked. Buttel offers several possible reasons for this. First, political economy, especially that with a neo-Marxist hue, has been somewhat overshadowed in recent decades by other trendy theoretical flavours, notably postmodernism and cultural studies. Second, treadmill theory has remained somewhat static, wedded to a manufacturing economy in a neo-liberal era in which Western economies seem to have shifted towards new information technologies, financial services and entertainment. Third, the notion of the treadmill is no longer very new or, in spite of what Schnaiberg believes, very controversial. As an analysis of industrial and consumer society the model now seems rather obvious, something that wasn't the case 30 years ago. Finally, the treadmill model fails to offer a viable green alternative to the capitalist economic system. Gould *et al.* (2004: 305) talk vaguely about an ecological synthesis that 'would extend the state's substantial control over ecosystems without regard to issues of profitability and wages/employment'. Alas, the treadmill theorists make no attempt to explain how this might be done in a capitalist society or whether a truly green capitalism is possible (Hannigan 2011: 49; Wright 2004).

Two normative theories of modernism and environmental improvement

When it comes to the environment, sociologists (and this applies equally well to other social scientists) commonly buy into two master narratives. One is the *apocalyptic narrative* wherein declining natural resources, increasing human population and runaway consumption converge, resulting in environmental collapse. As the Doomsday Clock indicates (see Chapter 1), we are now perched perilously at five minutes to midnight.

More optimistically, the *ecological enlightenment and sustainability narrative* detects a steady progress toward environmental awareness and improvement. It represents the latest chapter in an ongoing narrative of human progress that began with the Enlightenment. One broad manifestation of this is the *social learning* perspective, which is broadly inspired by Jurgen Habermas' notion of 'communicative action'. The core assumption here is that the environmental education efforts of scientists and social movements, nourished in a zone beyond the parameters of industrial society, will eventually begin to pay off and spur new ways of thinking and acting. That is, social learning involves a shift from the Dominant Social Paradigm (DSP) to the New Ecological Paradigm (NEP), much along the lines described by Catton and Dunlap.

Two specific analytic approaches that illustrate this ecological enlightenment and sustainability narrative stand out here: Mol and Spaargaren's 'ecological modernization' (EM) theory and Beck's 'risk society thesis'. Both are normatively charged, late modernist prescriptions. The two are frequently pitted against one another, insofar as the former is intended to transform economy–ecology contradictions into win–win situations, while the latter claims that our efforts to reform industrial society in the face of an apocalyptic eco-societal crisis are

Herculean (Blowers 1997; Desfor & Keil 2004: 62). Yet, both approaches share an important commonality: the expectation that an *environmental state* will eventually emerge, where environmental protection is accepted as a basic responsibility (Fisher 2003: 9–10).

Risk society thesis

The most influential attempt to update modernism has been Ulrich Beck's *risk society thesis*. In comparison to EM theory (see below), Beck is openly critical of modernity and its attendant risks. Nevertheless, he concludes that modernity ultimately has the capacity to solve the problems it produces (Barry 1999: 152).

Beck's thesis starts with the premise that Western nations have moved from an 'industrial' or 'class' society to a risk society in which the risks and hazards produced as part of modernization must be prevented, minimized, dramatized or channeled. Risk is said to be much more evenly distributed now than was formerly the case. As Beck memorably phrases it, 'hunger is hierarchical, smog is democratic'. Nevertheless, both the former 'wealth distributing society' and the emergent 'risk distributing society' contain inequalities.

Risk attached to events such as chemical spills and radiation poisoning are more than the unfortunate byproducts of industrialism and capitalism. Rather, they are a testament to the failure of social institutions, most notably science, to control new technologies. Such risks transcend both space and time. The 1986 Chernobyl nuclear accident in the Ukraine is a dramatic illustration of this. Due to the 'boomerang effect', risks that are exported abroad (such as agricultural herbicides and pesticides) inevitably come back to haunt us. Finally, risks today are said by Beck to be largely invisible to lay people, identifiable only through sophisticated scientific instrumentation.

One important feature of the risk society is the way in which the past monopoly of the sciences on rationality has been shattered. Paradoxically, science becomes 'more and more *necessary*, but at the same time, *less and less sufficient* for the socially binding definition of truth' (Beck 1992: 156). Beck contrasts the rigid 'scientific rationality' that prevailed for most of the twentieth century with a new 'social rationality' that is rooted in a critique of progress. Under pressure from an increasingly edgy public, new forms of 'alternative' and 'advocacy' science come into being and force an internal critique. This 'scientization of protest against science' produces a fresh variety of new public-oriented scientific experts who pioneer new fields of activity and application (e.g. conservation biology). In a similar fashion, monopolies on political action are said to be coming apart, thus opening up political decision-making to the process of collective action. One example of this is the entry of the Greens into parliamentary politics in Germany in the 1980s.

Finally, the dynamic of *reflexive modernization* leads to a greater individualization. Unbound from the strictures of traditional, pre-modern societies, the new urban citizens of the Industrial Revolution were expected to reach new levels of creativity and self-actualization. This did not happen, largely because a new constraint – the 'culture of scientism' – invaded every part of our lives from risk construction to sexual behaviour. Now there is an opportunity for individuals to once again break free and choose their own lifestyles, subcultures, social ties and identities (p. 131). Each of us, Beck believes, is obliged to reflect upon our personal experiences and make our own decisions about how we wish to live (Irwin 2001: 56). Ironically, just as the individualized private existence finally becomes possible, we are confronted with risk conflicts, which by virtue of their origin and design resist any individual treatment. Global environmental problems such as the greenhouse effect and the thinning of the ozone layer illustrate this. Thus the *reflexive scientization* in which scientific decision-making, especially that related to risk, is opened up to social rationality is vital to the reclamation of individual autonomy. Democracy should not, Beck insists, 'end at the laboratory door'.

While Beck's analysis is fresh and powerfully presented, it is not without problems. As Lidskog (1993) points out, Beck contradicts himself by arguing that the planet is in increasing peril due to an escalation of objectively certifiable global risks and, at the same time, insisting that risks are entirely socially constructed and therefore do not exist beyond our perception of them. Indeed, if you were to question Beck's assertion that the scope and effect of 'real' risk has sharply increased in late modernity, then this would have serious implications for the efficacy of his entire risk society thesis. Beck's response to this criticism is frustrating. He sees no contradiction between depicting a world in which risk is pervasive and possibly apocalyptic while observing that such risks are 'particularly open to social definition and construction' (1992: 23). Rather, he is preoccupied with promoting a distinctive vision of an ecologically rational or ecologically enlightened society (Barry 1999: 153).

Beck has also attracted considerable critical heat for his assertion that class-based rancor over the distribution of goods has fallen off in favour of new and shifting patterns of coalition and division. Increasingly, he ventures, it is not unusual to observe situations where workers in environmentally polluting industries join together with management in opposition to 'victims' from competing sectors of the economy such as fisheries and tourism. In some cases, alliances may even emerge between those once seriously in conflict with one another. For example, in New Mexico and Montana, ranchers and green organizations such as the Sierra Club have put aside their historic differences to jointly battle[7] against the common threat of proliferating oil and gas wells (Carlton 2005). This interpretation is flawed in that powerless economic actors are frequently compelled to support polluting technologies and policies in order to survive. Citing the case of Australian broadacre farmers who have come to accept chemical-dependent styles of agriculture as rational approaches to environmental management, Lockie (1997) notes that it is possible to be both a victim and a perpetrator at one and the same time. The farmer as perpetrator contributes to global pollution through engaging in chemical-intensive farming practices even as the farmer as victim is exposed to toxic materials that may be the source of chemically induced illness, ranging from headaches to cancer.

Finally, critics of the risk society thesis have accused Beck of being unacceptably vague about the details of political and scientific decision-making in the reflexive phase of modernity that he sees as hovering just around the corner. Seippel (2002: 215–6) implies that Beck's vision of politics in a 'civil society' is naïve and utopian. Why should we expect the political jockeying and dealing that are characteristic of traditional politics to suddenly disappear overnight? Indeed, in blurring the boundaries between conventional politics and civil society, we may even risk opening the latter up to undemocratic interests, values or modes of action. Furthermore, Beck overstates the potential for ecological rationality here, ignoring the cultural embeddedness of social interaction; there is little reason to expect that a society obsessed with celebrities and shopping will suddenly change direction and start making choices solely on the basis of new, post-materialist values. In short, as enlightening as it may seem, the risk society thesis ultimately constitutes a 'mythical discourse' (Alexander & Smith 1996).

Ecological modernization

Cast in the spirit of the Bruntland Report, ecological modernization, like sustainable development, indicates the possibility of overcoming the environmental crisis without leaving the path of modernization. In Huber's scheme, industrial society develops in three phases: (1) the industrial breakthrough, (2) the construction of industrial society and (3) the ecological switchover of the industrial system through the process of 'superindustrialization'. What makes this latter phase possible is a new technology, the invention and diffusion of microchip technology.

Ecological modernization rejects the Schumacher (1974)-inspired 'small is beautiful' ideology in favour of large-scale restructuring of production-consumption cycles to be accomplished through the use of new, sophisticated, clean technologies (Spaargaren & Mol 1992: 340). Unlike sustainable development, there is no attempt to address problems of the less developed countries of the South. Rather, the theory focuses on the economies of Western European nations, which are to be 'ecologized' through the substitution of microelectronics, gene technology and other clean production processes for older, 'end-of-pipe' technologies associated with the chemical and manufacturing industries. In contrast to Schnaiberg's treadmill of production perspective, capitalist relations of production, operating as a treadmill in the ongoing process of economic growth, are treated as largely irrelevant (Spaargaren & Mol 1992: 340–1).

According to Simonis (1989), the ecological modernization of industrial society contains three main strategic elements: a far-reaching conversion of the economy to harmonize it with ecological principles, a reorientation of environmental policy to the 'prevention principle' (seeking a better balance between stopping pollution before it happens and cleaning it up later on) and an ecological reorientation of environmental policy, especially by substituting statistical probability for 'prove-beyond-a-doubt' causality in legal suits against polluters. Unfortunately, little is said about the social and political barriers that need to be overcome in trying to implement these strategies, especially outside Germany and the Netherlands.

EM thinkers should be commended for staking out a reasoned position between catastrophic environmentalists, who preach that nothing less than total de-industrialization would suffice in saving the earth from an ecological Armageddon, and capital apologists who prefer a business-as-usual approach (Sutton 2004: 146). Alas, EM is hobbled by an unflappable sense of technological optimism.[8] All that is needed is to fast-forward from the polluting industrial society of the past to the new super-industrialized era of the future. Yet, the silicon chip revolution, which is the basis of this super-industrialization, is by no means environmentally neutral as the theory of ecological modernization suggests (Mahon 1985). According to the World Health Organization, exposure to lead, cadmium and mercury, all of which are in discarded cell phones, can cause irreversible neurological damage, especially in children. Furthermore, it is worth remembering that nuclear power was also touted as a clean, safe technology until its more undesirable features became known as a result of a series of mishaps, most recently the 2011 Fukushima nuclear accident in Japan.

As a sociological explanation, the theory of ecological modernization is as much prescriptive as analytic. Spaargaren and Mol initially said little about the power relations that characterize environmental processes, assuming that good sense must automatically triumph. Yet, sustainability, the guiding concept behind ecological modernization, is as much a political–economic dimension as an ecological one: what can be sustained is only what political and social forces in a particular historical alignment define as acceptable (Gould *et al.* 1993: 231).

Mol and Spaargaren have since presented a revisionist version of EM theory. The initial debates of the early 1980s, they caution, 'should be understood as an overreaction directed at the dominant schools of thought in environmental sociology and debate in the late 1970s and early 1980s' (2000: 18–9). Ecological modernization theory, they insist, was originally meant to challenge the notion put forward by both neo-Marxists and others that the modernization project was in its death throes; that the widespread environmental and ecological deterioration of the time was *prima facie* evidence of this; and that things could be salvaged only by fundamentally reshaping the core institutions of modern society.

Today, Mol and Spaargaren claim, these initial debates have become less relevant. Capitalism, they say, has evolved in a greener direction. For example, market-based instruments such as tradable pollution credits have displaced previous strategies that emphasized heavy-handed state regulation and enforcement. Furthermore, ecological modernization theorists themselves have incorporated critical comments from the earlier debate, reforming and refining their analysis of social change. They now claim to present a more nuanced position regarding capitalism, interpreting it 'neither as an essential precondition for, nor as the key obstruction to, stringent and radical environmental reform'.

Mol and Spaargaren place their faith in 'responsible capitalism' and the primacy of the market. In his empirical research into the Dutch chemical industry, a notorious polluter in the past, Mol (1997) finds nothing but good news. Reacting to consumer pressure, Dutch chemical companies initiated a spate of green measures, from the introduction of new technologies (low organic solvent paints) to new corporate instruments such as annual environmental reports, environmental audits and environmental certification systems. Together, Mol says, this represents 'a process of radical modernization' that has undercut any misguided 1970s and 1980s style demands for the dismantling of chemical production or even a shift to 'soft chemistry' (e.g. 'natural paints', which have failed to capture more than a one per cent share of the market in European countries). The institutions of modernity, Mol concludes, are by no means fading away; no massive movement away from a 'chemicalized' lifestyle can be identified and the erosion of trust in the scientific foundations of the chemical industry that might be inferred from Beck's risk society thesis is more or less absent.

Why do the treadmill analysts differ so broadly from the ecological modern-izationists? Schnaiberg suggests that it has to do with a difference in sampling approaches. EM theorists look at cutting edge corporate innovations or 'best practice' industries and assume that these changes will eventually diffuse widely. Treadmill theorists are skeptical, observing that the EM successes heralded by Mol and his colleagues may simply represent a 'creaming' of a programme of ecological incorporation into production practices (Schnaiberg *et al.* 2002: 29). In short, EM theorists are said to be naïve for claiming that greener production practices in arenas such as the Dutch chemical industry constitute a powerful 'third force' and part of a trajectory toward a future characterized by sustainability. Rather, firms that make ecological improvements do so either under direct pressure from state regulation or social movement action. Alternatively, these improvements are not real, having been achieved only through 'creative accounting' or misreporting (p. 29). Whether you regard environmental modernization as visionary or deluded is ultimately a measure of your degree of faith in gradualism as against the necessity of more radical solutions.

A major controversy: the realism versus constructionism debate

As Freudenburg (2000: 3) observes, 'more than any other subject in the discipline of environmental sociology, social construction[ism] has found fertile ground as well as fierce criticism'. Some analysts place constructionism at the very core of environmental theory. The idea that the environment is socially constructed, Lockie (2004: 29) notes, is 'perhaps one of the most fundamental concepts within environmental sociology'. Others reject this claim to being a full blown, coherent theory as 'exaggerated', arguing that, at best, it should be seen in more modest terms as 'a set of concepts and methodological conventions' (Buttel *et al.* 2002: 25). Less diplomatic critics depict the social constructionist as a sort of Darth Vader, perverting the force of sociological understanding and ignoring the reality of the environmental crisis. Noted conservation biologist Michael Soulé condemns social constructionism as an academic 'fad' whose rhetoric 'justifies further degradation of wildlands for the sake of economic development' and whose relativism 'can be just as destructive to nature as bulldozers and chainsaws' (Soulé & Lease 1995: xv; cited in Smith 1999: 362). In a similar pitch, environmental ethicist Eileen Crist (2004: 16) accuses constructionist analyses of nature of foolishly 'striving to interpret the world at an hour that is pressingly calling for us to change it'.

It is worth spending some time recalling why social constructionism first emerged as a way of dealing with environmental matters, what forms it assumed, why it has generated so much critical heat and how it might continue to make a useful contribution.

The case against constructionism

First and foremost, opponents of social constructionism object to what they perceive as its neutral stance in the midst of the looming environmental 'crisis' (see Chapter 1). As a popular saying from the 1960s put it, 'If you're not part of the solution, you're part of the pollution'.[9] Social constructionists are routinely pilloried for allegedly denying that the earth is under siege from a host of environmental hazards ranging from nuclear power leaks to anthropocentric global climate change. Brulle (1998: 138–9) rues the reluctance of constructionists to rank specific environmental issues as more serious and deserving of action than others. Failing to do so fundamentally undermines the legitimacy of the arguments that environmental problems are real and legitimate and thus deserve our attention for their resolution.

Williams (1998) agrees, calling the constructionist position inadequate 'because it leads to a relativizing perspective where no claim to reality is privileged over any other' (p. 478). He recommends that researchers select between competing constructions of environmental problems, since the consequences of not so doing are potentially profound. Williams accuses constructionists of lending tacit support to economic and political powerful interests by stressing that environmental threats are riddled with 'multiple and contradictory uncertainties'.[10] For example, he cites the actions of the Western Fuels Association, a US industry trade group, in reprinting and distributing articles that express uncertainty about specific scientific issues related to global warming. This is evidence, he says, that powerful social interest groups will exploit any weakness created by constructionist expressions of scientific uncertainty. By contrast, a 'realist' view asserts that 'the physical destruction of the environment can be empirically measured and scientifically monitored, thus avoiding an extreme form of naïve constructionism' (Picou & Gill 2000: 145).

Additionally, realists charge that the conflicting uncertainties approach adopted by constructionists privileges a contingent of 'rogue' scientists over the 'responsible' majority. For example, it is alleged that there is currently a unanimous scientific consensus that the earth is heating up and that this global climate shift is primarily due to humanly produced greenhouse gas emissions (Oreskes 2004). The small handful of scientists who dissent from this view, it is argued, are not legitimate because they are firmly 'in the pocket' of various corporations, state officials and anti-climate change interest groups who simply don't want to make the costly policy changes that would be required to comply with international accords such as the Kyoto Protocol (Buttel *et al.* 2002: 23). For opponents of Kyoto, the vital strategic task is allegedly to keep the public believing that there is no consensus about global warming in the scientific community; here, it is said, is where constructionists naïvely betray the environmental cause by encouraging this 'fiction'.

Constructionists reply

Social constructionists call this a grave misrepresentation. Only a 'false reductionism', Wynne (2002: 472) says, can construe constructionist accounts as claiming that environmental risks don't exist or that natural reality plays no identifiable role in producing knowledge about these risks. What constructionists are actually saying is that we need to look more closely at the social, political and cultural processes by which certain environmental conditions are defined as unacceptably risky, and therefore contributory to the creation of a perceived state of crisis.

Constructionists insist that bestowing *absolute* certainty solely on the basis of a scientific head count is perilous. After all, scientific consensus once dictated that the earth was flat and that the primary source of disease was 'vapours'. Yearley (2010: 214–5) points to the inconsistency of environmentalists, who invoke the power of the majority in the case of global warming, but contest official expert scientific advice in other controversial areas, notably genetically modified organisms.

In the case of global warming, the debate is by no means 'settled', as is frequently asserted. A prominent source for that assertion is an article in the top journal *Science*, 'Beyond the Ivory Tower', by social studies of science scholar turned global warming activist Naomi Oreskes. She conducted a qualitative analysis of 928 abstracts of papers on climate change published in scientific journals published between 1993 and 2004. Oreskes wrote:

> Such statements suggest that there might be substantive disagreement in the scientific community about the reality of anthropogenic climate change. This is not the case. The scientific consensus is clearly expressed in the reports of the Intergovernmental Panel on Climate Change (IPCC).
>
> (Oreskes 2004: 1686)

Those who cite Oreskes' study seem unaware of a 'repeated' (not the same as a 'panel' study) survey of climate scientists carried out in 1996, 2003 and 2008 by Dennis Bray and Hans von Storch. Rather than infer from the content of journal abstracts, they surveyed climate scientists directly, including authors who had contributed to Oreskes' study. Bray asks:

> But how should consensus be represented, for example should it be a simple level of agreement with all aspects of the reports, some aspects of the reports? One would expect that within any complex issue requiring multiple levels of expertise there would be matters where a consensus exists and matters where there is little or no consensus, or at least, some doubt.
>
> (Bray 2010)

To capture this complexity, in the 2008 survey the researchers devise three dimensions of consensus as it pertains to climate science: manifestation, attribution and legitimation. *Manifestation* asks how likely it is that global climate change is already underway. *Attribution* asks how likely it is that climate change is happening as a result of anthropocentric (human) influences. *Legitimation* asks about agreement with the IPCC reports. This final dimension is further divided into four parts. Respondents were asked whether the IPCC reports accurately reflect the consensus of scientific thought pertaining to (i) temperature, (ii) precipitation, (iii) seal level rise or (iv) extreme events. Finally, respondents were asked about their participation or non-participation in formulating the IPCC reports and sub-divided accordingly.

Bray reports that roughly one in five respondents (16 per cent for IPCC participants, 24 per cent for non-participants) claim that the current state of scientific knowledge is not developed well enough to allow for a reasonable assessment of the effects of greenhouse gases emitted from anthropogenic sources. In similar fashion, on the legitimation dimension, a significant minority of respondents questioned the accuracy of the IPCC report. While they were more likely to indicate that the IPCC 'underestimated' the threat of global climate change, nevertheless, about 15–20 per cent of the sample said that the 'state-of-the-art' IPCC document 'overestimated' the risk. Among non-IPCC participants, this ranged from a high of 21 per cent for temperature to a low of 12 per cent for sea level rise.

Bray concludes, 'What this analysis has disclosed is that the science is NOT settled and that perhaps beneficial scientific skepticism, albeit in an infant stage, is growing' (2010: 350). It must be noted that that Bray and von Storch *only* sent their questionnaire out to climate scientists. Not included in the survey were geoscientists, a scientific specialty group who might be expected to be far less accepting of the IPCC version of climate change.

Health and environmental threats do not always follow a unidirectional path.

Consider, for example, the so-called 'killer obesity' crisis. In the wake of a high profile article in the *Journal of the American Medical Association* (9 March, 2004) reporting that 'obesity' had caused 400,000 deaths in 2000, up 33 per cent from a decade earlier, poor eating habits were confirmed as a major preventable killer. A year later, the study's authors, the US Centers for Disease Control, corrected these figures, downsizing the number of obesity deaths to 26,000 and revealing that 86,000 moderately overweight Americans were actually found to have *lived longer* than people of normal weight (Henninger 2005). By mid-2005, the pendulum had begun to swing back, as indicated by an article in *Scientific American* entitled 'Obesity: An Overblown Epidemic?' (Gibbs 2005).

This doesn't mean that we should not worry about the alarming incidence of obesity rates, especially among children; nor should we relax our concerns about the possibility of the polar ice caps melting in the foreseeable future. What it suggests is that it is not wise to allow a *discussable issue* to become an *evident crisis*, especially where the evidence is open to multiple interpretations.

As we saw in the previous chapter, the first generation of environmental sociologists uncritically accepted the existence of an environmental crisis brought on by unchecked population growth, over-production and the adoption of dangerous new technologies. Thus, Dunlap and Catton's NEP, that 'provided the template for modern environmental sociology' (Buttel 2000: 19) is basically an analogue for the ecocentric claims of radical ecologists that nature must be placed 'at the centre of moral concern, politics and scientific study' (Sutton 2004: 78). Buttel *et al.* (2002: 22) point out that: 'prior to the late 1980s, a sizable share of the North American environmental sociology community saw its mission as being to bring the ecological sciences and their insights to the attention of the larger sociological community'. It was dominated by an environmental realism that was 'driven by the impulse of "saving the Earth", pointing to the ongoing environmental destruction and a future global catastrophe' (Lidskog 2001: 120). In this context, constructionism is treated as a 'spoiler'.

While not denying the validity of concern over pollution, energy shortages and runaway technology, social constructionists insist that the central task ahead for environmental sociologists is *not* to document these problems but rather to demonstrate that they are the products of a dynamic social process of definition, negotiation and legitimation. As Yearley (1992: 186) notes, demonstrating that an environmental problem has been socially constructed is not to undermine and debunk it, since 'both valid and invalid social problem claims have to be constructed'. Similarly, as Dryzek points out,

> Just because something is socially interpreted does not mean it is unreal. Pollution does cause illness, species do become extinct, ecosystems cannot absorb stress indefinitely, tropical forests are disappearing. But people can make very different things of these phenomena and – especially – their interconnections, providing grist for political dispute.
>
> (2005: 12)

In short, social constructionism does not deny the considerable powers of nature. Rather, it asserts that the magnitude and manner of this impact is open to human construction.

Constructionists point out that the rank ordering of environmental problem claims by social actors does not always correspond to actual need; rather, it reflects the political nature of agenda setting. Thus, Yearley concludes:

> There are good grounds for believing that the topics that rise to the top of the public's attention are not those where the reality of the problem is most well documented or where the real impacts are the greatest, but those where the agents that propel issues into the public consciousness have worked the most effectively.
>
> (2002: 276)

Constructionists insist any claim can be evaluated on the basis of hard evidence such as statistics or public opinion polls, even if these are in themselves social constructions (Best 1989: 247). The researcher is encouraged to consider the historical context within which the claim has been formulated in order to explain the emergence and assess its validity (Rafter 1992). Agnosticism does not mean that we must automatically accord all claims equal validity. We might reasonably doubt the widely publicized media claims by the Raelians, a flying saucer cult, that they have successfully cloned several humans in their laboratory. On the other hand, a warning from prominent public health officials that tens of millions of urban residents could become clinically ill during a potential outbreak of a global influenza pandemic next winter would carry more authority, even if it isn't a certainty. As Jones (2002: 249) points out, the social constructionist approach 'does not have to lead to a relativism wherein there are no reasonable grounds to choose between different scientific knowledges or environmental histories'.

Finally, constructionists point out that their approach is more consistent with the existing canon of sociological theorizing than are more explicitly 'ecological' approaches. Greider and Garkovitch (1994) argue that the role of the environmental sociologist should lie not in a quest for some elusive new model that causally links ecosystem breakdown with social variables (see Catton 1994) but in a return to classic sociological questions of perception and power. In this context, biophysical changes in the environment are meaningful only insofar as groups affected by these changes come to acknowledge them through a self-redefinition. For example, in addressing the political conflict in the American Northwest over the spotted owl, the key question for the sociological analysts should not be the number of owls but the way in which the fluctuating power of the different social actors or claims-makers – loggers, rural businesses, international logging companies, environmentalists – shapes the definition of the situation (Greider & Garkovitch 1994: 21). Greider and Garkovitch treat global environmental change as a type of 'landscape'. In looking at how landscapes such as this are symbolically created and contested, researchers are contributing to the furtherance of a well-established school of thought in sociology and helping to forge a role for the discipline in the debates over environmental issues.

Prospects for reconciliation

The constructionist–realist 'war' has recently begun to settle, with proponents and opponents alike acknowledging that these sharp exchanges have become repetitive and counterproductive. Even so, the debate hasn't vanished altogether. Dunlap (2010: 16) discerns a broader cleavage between constructivist-oriented scholars committed to 'environmental agnosticism' and realist-oriented scholars practicing 'environmental pragmatism', reflecting in part the different outlooks of European and North American environmental sociologists. Dunlap, who readily acknowledges a preference for environmental pragmatism versus agnosticism, nevertheless hopes to see greater efforts to merge the strength of the two approaches, 'with agnostics using their rich analytical tools to delve more deeply into the material world and pragmatists paying greater attention to the impact of constructions, values, culture and the like' (Dunlap, 2010: 28).

In her review of the first edition of *Environmental Sociology*, Katherine Betts (1997) argues that the ecological paradigm, as typified by Catton and Dunlap's work, can potentially accommodate the sociology of knowledge approach preferred by constructionists. On their part, realist scholars must take care not to use the truth claims of these ideas as ideologies as a sufficient explanation for their success or failure. Constructionists, on the other hand, need to abandon their stance of studied indifference to substantive environmental questions. One way to do this is to acknowledge that agnosticism in one particular aspect of their work does not commit them to agnosticism in all aspects of their work. In so doing, the environmental sociologist can avoid having to choose between disinterested scholarship and committed activism.

Conclusion

As a general rule, new academic specialities develop in tandem with the emergence of new movements for change in the wider society. Although there had previously been clusters of sociological writing on natural resources, population and economy, environmental sociology as a distinct area of interest didn't appear until the 1970s, in response to the rising profile of environmentalism and the environmental movement. Classic sociology texts by Durkheim, Marx and Weber were of minimal assistance here. Initially, two explanations for the contemporary spike in environmental degradation and destruction took centre court: the 'ecological' explanation associated with Catton and Dunlap, and the 'political economy' explanation, especially Schnaiberg's theory of the 'societal-environmental dialectic' and the 'treadmill of production'. In its second wave of growth as a sociological area of enquiry, environmental sociology concerned itself more with issues of environmental improvement and the state. Two 'normative' perspectives stand out here: Ulrich Beck's 'risk society' thesis and 'ecological modernization' theory, as elaborated by Mol and Spaargaren. In the 1980s and 1990s, a major controversy raged in environmental sociology between the 'realism' and 'social construction-ism'. More recently, however, leading figures from both 'camps' have softened their stance and worked towards forging an accommodation.

Notes

1 Addressing this point directly, Sorokin comments, 'A reader of these lines may think Dr. Huntington has at his disposal there the detailed record of the Meteorological Bureau of Ancient Rome' (1964) [1928: 191].

2 Dickens (1992) suggests that structural functionalists, especially their dean, Talcott Parsons, might have gone further and actually developed a theory of social evolution in an environmental context that stresses how biological inheritance permitted humans to both adapt to the natural world and to change it. Alas, this potential was never developed, leaving environmental factors as marginal elements in sociological explanation.

3 This view is not, however, universally shared. For example, Goldblatt (1996: 3) states 'of the classical trinity [Durkheim, Weber, Marx], Weber's work conducts the most limited engagement with the natural world'.

4 One more recent effort along these lines is York et al.'s (2003: 36–8) discussion of how the metabolic rift can lead to increases in GHG (greenhouse gas) emissions. Three ways this occurs are specified: the increased transportation of natural resources necessitated by urbanization, the replacement of organic matter by chemical fertilizers and the diversion of methane-generating organic waste to landfills rather than back into the soil.

5 I am grateful to Filip Alexandrescu for this insight and the references from Boulding's work in the 1950s.

6 This encompasses Schnaiberg and his former doctoral students, Kenneth Gould, David Pellow and Adam Weinberg.

7 This opposition evidently did not extend to shale gas drilling (fracking), where the Sierra Club, under the umbrella of controlling carbon emissions, briefly joined forces with a leading energy producer to battle the coal mining industry (see Chapter 9).

8 Spaargaren (2000: 64–5) objects to this statement, declaring the optimism/pessimism dichotomy to be less than helpful. Science and technology, he argues, are important vehicles in the ecological modernization process, but this 'does not imply, however that one would automatically or inevitably lapse into a technological fix approach'. This is more the case for a 'strong ecological modernisation', which purports to be more open to broad-ranging changes to society's institutional structure and economic system than for 'weak ecological modernisation' that emphasizes technological

solutions to environmental problems and 'looks like a discourse for engineers and accountants' (Dryzek 2005: 172–3).

9 For example, New York governor Nelson Rockefeller used this slogan during a speech in Prospect Park, Brooklyn, on Earth Day, 1970 (Grayson 2006).

10 As Thompson (1991) notes, environmental debates reflect the existence not just of an absence of certainty but rather of *contradictory certainties*: several divergent and mutually irreconcilable sets of convictions both about the difficulties we face and the available solutions.

Further reading

Beck, U. (1992) *Risk Society: Towards a New Modernity*, London: Sage.

Davidson, D.J. and Gismondi, M. (2011) *Challenging Legitimacy at the Precipice of Energy Calamity*, New York: Springer.

Gould, K.A., Pellow, D.N. and Schnaiberg, A. (2008) *The Treadmill of Production: Injustice and Unsustainability in the Global Economy*, Boulder, CO: Paradigm Publishers.

Murphy, R. (1994) *Rationality and Nature*. Boulder, CO: Westview.

Woodgate, G. and Redclift, M. (eds) (2010) *The International Handbook of Environmental Sociology (2nd edition)*, London: Edward Elgar.

Online resources

American Sociological Association, Section on Environment and Technology. Available HTTP: http://www.envirosoc.org

International Sociological Association, Research Committee on Environment and Society (RC24). Available HTTP: http://www.environment-societyisa.org

3 Social construction of environmental issues and problems

In his contribution to a recent text on global environmental politics, Paul Harris states: 'The reason that climate change has found its way onto the international agenda is primarily because its causes and consequences have become so evident and prudentially important' (Harris 2011: 114). This flies in the face of the primary orientation of this book. As Rice (2013: 239–40) recognizes, 'Environmental concern [Hannigan (1995) further noted] is not constant but fluctuating, and environmental problems do not necessarily become imprinted on individual and societal consciousness due to worsening objective conditions'. Rather, their progress varies in direct response to successful *claims making* and contestation by a cast of social actors that includes scientists, industrialists, politicians, civil servants, journalists and environmental activists. In the case of China's recent toxic carbon cloud (Box 3.1), hard-core pollution from coal-fired energy plants had been around for quite some time, but only became a national emergency when environmentalists, Internet bloggers and the media labeled it as the nation's *airpocalypse*.

Environmental problems are similar to social problems in general. There are, however, a few notable differences. While social problems frequently cross over from a medical discourse to the arenas of public discourse and action (Rittenhouse 1991: 412), they nevertheless derive much of their rhetorical power from moral rather than factual argument. By contrast, environmental problems such as pesticide poisoning or global warming are tied more directly to scientific findings and claims (Yearley 1992: 117). This is true even in the case of environmental justice claims, which are among the most morally charged indictments of corporate and state polluters. Furthermore, although they are traceable to human agents, environmental problems have a more imposing physical basis than social problems, which are more rooted in personal troubles that become converted into public issues (Mills 1959).

The constructionist interpretation that I present in this chapter is traceable to a paradigm shift that transformed the sociology of social problems in the early 1970s.

Constructing social problems

Nearly a quarter of a century ago, the sociology of social problems first began to experience a major conflict with the appearance of a seminal article by Malcolm Spector and John Kitsuse (1973) entitled 'Social Problems: A Reformulation'. Here, and in a subsequent book (1977), Spector and Kitsuse challenged the structural functional approach to social problems that had theretofore dominated the field. Functionalism, as exemplified by the work of Merton and Nisbet (1971),

> assumed the existence of social problems (crime, divorce, mental illness) which were the direct products of readily identifiable, distinctive and visible objective conditions. Sociologists were regarded as experts who employ scientific methods to locate and analyze these moral violations and advise policymakers on how best to cope. In addition, the sociologist's role was to bring to lay audiences an awareness and understanding of worrisome conditions, especially where these were not readily evident.
>
> (Gusfield 1984: 39)

Spector and Kitsuse argue that social problems are not static conditions but rather 'sequences of events' that develop on the basis of collective definitions. Accordingly, they define social problems as 'the activities of groups making assertions of grievances and claims to organizations, agencies and institutions about some putative conditions' (1973: 146). From this point of view, the process of claims-making is treated as more important than the task of assessing whether these claims are truly valid or not. For example, rather than document a rising crime rate, the social problems analyst is urged to focus on how this problem is generated and sustained by the activities of complaining groups and institutional responses to them (1973: 158). Since 1973, *social constructionism* has increasingly moved towards the core of social theorizing, generating a critical mass of theoretical and empirical contributions both within the social problems area and across sociology as a whole.

Constructionism as an analytic tool

Best (1989b: 250) has noted that constructionism is not only helpful as a theoretical stance but it can also be useful as an analytic tool. In this regard, he suggests three primary foci for studying social problems from a social constructionist perspective: the claims themselves, the claims-makers and the claims-making process.

Nature of claims

As initially conceptualized by Spector and Kitsuse, claims are complaints about social conditions that members of a group perceive to be offensive and undesirable.

According to Best (1989b: 250), there are several key questions to be considered when analyzing the content of a claim: what is being said about the problem? How is the problem being typified? What is the rhetoric of claims-making? How are claims presented so as to persuade their audiences? Of these, it is the third question that has generated the most interest among contemporary social problems analysts. Using the example of the 'missing children' i.e. runaways, child-snatched abductions by strangers, Best (1987) analyzes the content of social problems claims by focusing on the 'rhetoric' of claims-making. Rhetoric involves the deliberate use of language in order to persuade. Rhetorical statements contain three principal components or categories of statements: grounds, warrants and conclusions.

Grounds or data furnish the basic facts that shape the ensuing policy-making discourse. There are three main types of grounds statements: definitions, examples and numeric estimates. Definitions set the boundaries or domain of the problem and give it an orientation; that is, a guide to how we interpret it. Examples make it easier for public bodies to identify with the people affected by the problem, especially where they are seen as helpless victims. Atrocity tales are one especially effective type of example. By estimating the magnitude of the problem, claims-makers establish its importance, its potential for growth and its range (often of epidemic proportions).

Warrants are justifications for demanding that action be taken. These can include presenting the victim as blameless or innocent, emphasizing links with the historical past or linking the claims to basic rights and freedoms. In analyzing the professional literature on 'elder abuse', Baumann (1989) identified six primary warrants: (1) the elderly are dependent; (2) the elderly are vulnerable; (3) abuse is life-threatening; (4) the elderly are incompetent; (5) ageing stresses families; (6) elder abuse often indicates other family problems. Hannigan (2012a: 84–5) identifies two warrant 'bundles' or 'packages' that have been deployed to make the case for engaging in climate change adaptation and disaster risk reduction. In the first warrant package, climate change is framed as a 'development emergency', undermining global efforts towards achieving sustainable development and poverty eradication. In the second bundle of warrants, which is framed in terms of social justice and environmental equity, climate change is depicted as magnifying the already uneven global distribution of risk, especially when it comes to poor communities situated at low elevations.

Conclusions spell out the action that is needed to alleviate or eradicate a social problem. This frequently entails the formulation of new social control policies by existing bureaucratic institutions or the creation of new agencies to carry out these policies.

Best further proposes two rhetorical themes or tactics that vary according to the nature of the target audience. The *rhetoric of rectitude* (values or morality require that a problem receive attention) is most effective early on in a claims-making campaign when audiences are more polarized, activists are less experienced and the primary demand is for a problem to be viewed in a new way. By contrast, the *rhetoric of rationality* (ratifying a claim will earn the audience some type of concrete benefit) works best at the later stages of social problems construction when claims-makers are more sophisticated, the primary demand is for detailed policy agendas and audiences are more persuadable. Rafter (1992: 27) has added another rhetorical tactic to Best's list: that of archetype formation. *Archetypes* are the templates from which stereotypes are minted and therefore possess considerable persuasive power as part of a claims-making campaign.

A further set of rhetorical strategies in claims-making has been proposed by Ibarra and Kitsuse (1993), who outline a variety of rhetorical idioms, motifs and claims-making styles.[1]

Rhetorical idioms are image clusters that endow claims with moral significance. They include a 'rhetoric of loss' (of innocence, nature, culture, etc.); a 'rhetoric of unreason' that invokes images of manipulation and conspiracy; a 'rhetoric of calamity' (in a world full of deteriorating conditions, epidemic proportions are claimed for a few; for example, AIDS or the 'greenhouse effect'); a 'rhetoric of entitlement' (justice and fair play demand that the condition, or as Ibarra and Kitsuse term it, the 'condition-category', be redressed) and the 'rhetoric of endangerment' (condition-categories pose intolerable risks to one's health or safety).

Rhetorical motifs are recurrent metaphors and other figures of speech (AIDS as a 'plague', the depletion of the ozone layer as a 'ticking time bomb') that highlight some aspect of a social problem and imbue it with a moral significance. Some motifs refer to moral agents, others to practices and still others to magnitudes (Ibarra & Kitsuse 1993: 47).

Claims-making styles refer to the fashioning of a claim so that it is in sync with the intended audience (public bodies, bureaucrats, etc.). Examples of claims-making styles include a scientific style, a comic style, a theatrical style, a civic style, a legalistic style and a subcultural style. Claims-makers must match the right style to the right situation and audience.

Claims-makers

In looking at the identity of claims-makers, Best (1989b: 250) advises that we pose a number of questions. Are claims-makers affiliated to specific organizations, social movements, professions or interest groups? Do they represent their own interests or those of third parties? Are they experienced or novices (as we have seen, this can influence the choice of rhetorical tactics)?

Many studies that have been undertaken in the social constructionist mode have pointed to the important role played by medical professionals and scientists in constructing social problems claims. Others have noted the importance of policy or issue entrepreneurs: politicians; public interest law firms; civil servants whose careers are dependent upon creating new opportunities, programmes and sources of funding. Claims-makers may also reside in the mass media, especially since the manufacture of news depends upon journalists, editors and producers constantly finding new trends, fashions and issues.

The cast of claims-makers who combine to promote a social problem can be quite diverse. For example, Kitsuse *et al.* (1984) identify three main categories of claims-makers in the identification of the *kikokushijo* problem in Japan, i.e. the educational disadvantage of Japanese schoolchildren whose parents have taken them abroad as part of a corporate or diplomatic posting: officials in prestigious and influential government agencies; informally organized groups of diplomatic and corporate wives; and the '*meta*' – a support group of young adults who have been victims of the *kikokushijo* experience.

It is also important to keep in mind that not all claims-makers are to be found among the grassroots or civil society. For example, it has been suggested that the contemporary 'obesity crisis' has been captained by 'a relatively small group of scientists and doctors, many directly funded by the weight-loss industry, [who] have created an arbitrary and unscientific definition of overweight and obesity' (Oliver 2005; cited in Gibbs 2005: 72).

Claims-making process

Wiener (1981) has depicted the collective definition of social problems as a continually ricocheting interaction among three sub-processes: *animating the problem* (establishing turf rights, developing constituencies, funnelling advice and imparting skills and information); *legitimating the problem* (borrowing expertise and prestige, redefining its scope, e.g. from a moral to a legal question, building respectability, maintaining a separate identity) and *demonstrating the problem* (competing for attention, combining for strength, i.e. forging alliances with other claims-makers, selecting supportive data, convincing opposing ideologists, enlarging the bounds of responsibility). These are overlapping rather than sequential processes, which together result in a public arena being built around a social problem.

Hilgartner and Bosk (1988) have identified these *arenas of public discourse* as the prime location for the evaluation of social problem definitions. However, rather than examining the stages of problem development, they propose a model that stresses the competition among potential social problems for attention, legitimacy and societal resources. Claims-makers or 'operatives' are said to deliberately adapt their social problem claims to fit their target environments; for instance, by packaging their claims in a novel, dramatic and succinct form or by framing claims in politically acceptable rhetoric.

Best (1989b: 251) poses a number of useful questions about the claims-making process. Whom did the claims-makers address? Were other claims-makers presenting rival claims? What concerns and interests did the claims-makers' audience bring to the issue, and how did these come to shape the audience's responses to the claims? How did the nature of the claims or the identity of the claims-makers affect the audience's response?

Table 3.1 Key tasks in constructing environmental problems

	Task		
	Assembling	*Presenting*	*Contesting*
Primary activities	discovering the problem	commanding attention	invoking action
	naming the problem	legitimating the claim	mobilizing support
	determining the basis of the claim		defending ownership
	establishing parameters		
Central forum	science	mass media	politics
Predominant layer of proof	scientific	moral	legal
Predominant scientific role(s)	trend spotter	communicator	applied policy analyst
Potential pitfalls	lack of clarity	low visibility	co-optation
	ambiguity	declining novelty	issue fatigue
	conflicting scientific evidence		countervailing claims
Strategies for success	creating an experiential focus	linkage to popular issues and causes	networking
	streamlining knowledge claims	use of dramatic verbal and visual imagery	developing technical expertise
	scientific division of labour	rhetorical tactics and strategies	opening policy windows

Source: Author.

Key tasks/processes in the social construction of environmental problems

Defining environmental problems

In defining environmental problems, bringing them to society's attention and provoking action, claims-makers must engage in a variety of activities. Some of these are centrally concerned with the collective definition of potential problems, others with the collective action necessary to ameliorate them (Cracknell 1993: 4). This is not to say that elements of definition and action do not interweave constantly. Nevertheless, environmental problems follow a certain temporal order of development as they progress from initial discovery to policy implementation.

In this section of the chapter, I identify three central tasks that characterize the construction of environmental problems. In doing so, I draw upon two[2] prior models: Carolyn Wiener's (1981) three processes through which a public arena is built around a social problem, and William Solesbury's (1976) three tasks that are necessary for an environmental issue to originate, develop and grow powerful within the political system. In *The Politics of Alcoholism*, Wiener depicts the collective definition of social problems as a continuing ricocheting interaction among three processes: animating, legitimizing and demonstrating the problem. These are presented as overlapping rather than sequential processes; that is, they interact with one another rather than operate independently.

Solesbury's scheme is more concerned with the political fate of environmental concerns. He notes the 'continuing change in the agenda of environmental issues' that may be partly accounted for by changes in the state of the environment itself (Ungar 1992) and partly through changing public views as to which issues are important and which are not. All environmental issues, he states, must pass three separate tests: commanding attention, claiming legitimacy and invoking action. Like Wiener, Solesbury points out that these tasks may be pursued simultaneously in no particular order (Cracknell 1993: 5), although it would presumably be difficult to invoke policy changes before the problem is recognized and legitimized.

In considering the social construction of environmental problems, it is possible to identify three key tasks: assembling, presenting and contesting claims.

Assembling environmental claims

The task of assembling environmental claims concerns the initial discovery and elaboration of an incipient problem. At this stage, it is necessary to engage in a variety of specific activities: naming the problem, distinguishing it from other similar or more encompassing problems, determining the scientific, technical, moral or legal basis of the claim and gauging who is responsible for taking ameliorative action.

Environmental problems frequently originate in the realm of science. One reason for this is that ordinary people have neither the expertise nor the resources to find new problems. For example, knowledge about the ozone layer is not tied to our everyday experience; it is available only through the use of high technology probes into the atmosphere directly above the polar regions (Yearley 1992: 116).

Some problems, however, relate more closely to our life experiences. Concern over toxic wastes frequently begin with local citizens who come to draw a causal link between seeping dump sites and a perceived increase in the neighbourhood incidence of leukaemia, miscarriages, birth defects and other ailments. This is what occurred in Niagara Falls, New York State, where Lois Gibbs and her neighbours were the first to associate their health-related problems with the chemical wastes buried 30 years before in the abandoned Love Canal. Those whose jobs or recreational pursuits bring them into close contact with nature on a daily basis (farmers, anglers, wildlife officers) may also be the initial source of claims because they pick up early environmental warning signals such as reproductive problems in livestock or mutations in fish. Acid rain was first launched as a contemporary environmental problem when a fisheries inspector in a remote area of Sweden telephoned researcher Svante Oden with the observation that there appeared to be a link between a rising incidence of fish deaths and an elevation in the acidity of lakes and rivers in the area.

Practical knowledge about the environment often originates from the everyday experience of villagers, small farmers and others in Southern societies. Sir Albert Howard, often regarded as the originator of organic agriculture, derived many of his ideas from consulting with peasant cultivators in India whom he called his 'professors' (Howard 1953): a strategy that was considered revolutionary in the context of British colonial administration. More recently, grassroots activists in Southern countries have emphasized the importance of 'ordinary knowledge' (Lindblom & Cohen 1979) that depends more on keen observation and common sense than on professional techniques. This ordinary knowledge is accumulated within local grassroots networks by breathing air, drinking water, tilling soil, harvesting forest produce and fishing rivers, lakes and oceans (Breyman 1993: 131). In a similar fashion, native (aboriginal) people in Northern societies accumulate firsthand knowledge of the environment that may not be available to non-indigenous observers. For example, it has been suggested[3] that biologists estimating the effect of mega projects on the ecology of rivers in the Canadian north may overlook the existence of a number of fish species simply because they never bother to ask native residents who know the land intimately (Richardson *et al.* 1993: 87).

A Swedish researcher, Karin Gustafsson (2011), assigns a greater role to local residents in the construction of environmental problems than is commonly acknowledged. Too often, she says, this process is depicted as if all environmental problems are exclusively global problems. Even where locals alert scientists to worrisome environmental conditions, this is soon co-opted by experts, who seal it within a global sphere. Drawing upon a case study of a pine-devouring moth invasion on the island of Gotland in the Baltic Sea, Gustafsson demonstrates how local residents' narratives played a crucial role in constructing the problem. In so doing, they 'transform the scientific knowledge in ways that made it possible to combine it with local knowledge into trustworthy claims' (p. 667).

In a similar fashion, Peuhkuri (2002: 159) observes that we should not assume that local lay knowledge is inevitably a counter pole to science in environmental conflicts. In contemporary societies, where the media and education penetrate even peripheral regions, 'local knowledge is a mixture of traditional knowledge, knowledge based on the local people's own observations and popularized science'.

In researching the origins of environmental claims, it is important for the researcher to ask where a claim comes from, who owns or manages it, what economic and political interests claims-makers represent and what type of resources they bring to the claims-making process.

In the early US conservation movement, environmental claims were largely traceable to an East Coast elite who utilized a network of 'old boy' ties to secure funding and political action. Enthusiastic amateurs, they dominated the boards of zoos, natural history museums and other public institutions from whence they were able to direct campaigns to save redwood trees, migratory birds, the American bison and other endangered species and habitats (Fox 1981). In a similar fashion, the threat to British birds, wildlife sites and other elements of nature was proclaimed in the late nineteenth and early twentieth centuries by a number of conservation groups with elite membership (Evans 1992; Sheail 1976).

By contrast, present day environmental claims-makers are more likely to take the form of professional social movements with paid administrative and research staffs, sophisticated fund-raising programmes and strong, institutionalized links both to legislators and the mass media. Some groups even use door-to-door canvassers or street corner solicitors who are paid an hourly wage or get to keep a percentage of their solicitations. Campaigns are planned in advance, often in pseudo-military fashion. Grassroots participation is not encouraged beyond 'paper memberships' with control centralized in the hands of a core group of full-time activists.

The process of assembling an environmental claim often involves a rough division of labour. While there are notable exceptions, research scientists are normally handicapped by a combination of scholarly caution, excessive use of technical jargon and inexperience in handling the media. As a result, an important finding may lie fallow for decades until proactively transformed into a claim by entrepreneurial organizations (Greenpeace, Friends of the Earth, Sierra Club) or individuals (Paul Ehrlich, Jeremy Rifkin, Bill McKibben). Greenpeace's claims-making activity does not so much flow out of its ability to construct entirely new environmental problems but rather from its genius in selecting, framing and elaborating scientific interpretations that might otherwise have gone unnoticed or been deliberately glossed over (Hansen 1993b: 171). Indeed, the nature of the relationship between the news media and environmental pressure groups such as Greenpeace has become sufficiently institutionalized (Anderson 1993a: 55) that it would be difficult for an emergent problem to penetrate the mass media arena without at least token validation from the latter.

In assembling an environmental problem, not all explanations are created equally. Claims that hinge on difficult to understand concepts such as 'entropy' are far less likely to stick than those that have at their nucleus more readily comprehensible constructs; for example, 'extinction' or 'overpopulation'. Sometimes, the basic outline of a claim only becomes clear in the context of a political, economic or geographic crisis. This was the case in 1973 when concerted action by OPEC (Organization of Petroleum Exporting Countries), the oil producers' cartel, triggered an energy crisis in industrial nations in the West.

Similarly, the abnormally hot US summer of 1988 gave the problem of global warming a visible, experiential focus. Most recently, the storm surge accompanying Hurricane Sandy in October, 2012 caused unprecedented destruction along the Jersey Shore and Long Island, as well as in Manhattan where it closed the subways. All of a sudden, New York's vulnerability to flooding and the possible role played by climate change became major topics of conversation.

Presenting environmental claims

In presenting an environmental claim, issue entrepreneurs have a dual mandate: they need both to command attention and to legitimate their claim (Solesbury 1976). While not unrelated, these constitute two quite separate tasks.

As Hilgartner and Bosk's (1988) model emphasizes, the arenas through which social problems become defined and conveyed to the public are highly competitive. To command attention, a potential environmental problem must be seen to be novel, important and understandable – the same values that characterize news selection in general (Gans 1979).

One effective way of commanding attention is through the claimants' use of evocative verbal and visual imagery. The extreme thinning of the ozone layer became much more saleable as an environmental problem when depicted as an expanding 'hole'; American children's entertainer, Bill Shontz has even recorded a hit song entitled *Hole in the Ozone*. Similarly, the effects of acid rain were successfully dramatized when German environmentalists began to use the term *waldsterben* (forest die-back). Larson *et al.* (2005) have demonstrated the prevalence of militaristic metaphors (attack, destroy, wipe out, contain, counteroffensive, full-scale war) in the media reporting of three contested areas of science-society discourse (invasive species, foot and mouth disease and SARS [Severe Acute Respiratory Syndrome]). Visual language can be especially powerful in carrying out this task. For example, technical data on the size of seal herds and codfish stocks instantly lost relevance when Brian Davies and other activists released photos to the media of baby seal pups being clubbed to death on the ice floes of Labrador.

It is not unusual, however, for these visual images to be streamlined so as to underline a central image. Mazur and Lee (1993: 711) give several striking examples of this. The NASA satellite pictures of the ozone hole over the Antarctic, which became a 'logo' of the problem, transformed continuous gradations in real ozone concentration into an ordinal scale that is colour-coded, conveying the erroneous impression that a discrete, identifiable hole could actually be located in the atmosphere over the South Pole. In August 1988, a *New York Times* article on rainforest destruction was accompanied by a stunning satellite photograph of the burning Amazon created by Alberto Setzer of the Brazilian Institute of Space Research. The photograph showed what appeared to be nearly 100,000 fires; however, it was really a composite of many separate pictures and included fires in areas of secondary forest growth as well as virgin rainforest. Rajão and Vurdubakis (2013) report that the use of satellite images to detect 'illegal deforestation' in the Amazon escalated in the 1980s as a result of growing national and international pressures aggravated by Space Shuttle pictures of huge fires consuming the rainforest. Federal rangers were tasked with inspecting forest clearings so as to establish a *correspondence* with satellite images. In this way, a phenomenon that is somewhat fuzzy and ambiguous is transformed into a straightforward fact, fixed in time and space.

Environmental issues may be forced into prominence when exemplified by particular incidents or events, for example, the nuclear accidents at Chernobyl and Three Mile Island, the Bhopal chemical disaster and the wreck of the oil tankers *Torrey Canyon* and *Exxon Valdez*. Dramatic events such as these are important because they assist political identification of the nature of an issue, the situations out of which it arises, the causes and effects, the identity of the activities and the groups in the community that are involved with the issue (Solesbury 1976: 384–5).

Staggenborg (1993) has identified six major types of 'critical events' that affect social movements such as the environmental movement. Large-scale socio-economic and political events such as wars, depressions and national elections influence the opportunities for collective action by altering perceptions of grievances and threats; for example, the 1980 election of US president Ronald Reagan led to increased memberships in environmental groups[4] since it raised the spectre of a free enterprise run rampant in national parks and other wilderness settings. National disasters and epidemics can represent a turning point in the movement, highlighting grievances and bringing about movement growth. Similarly, industrial and nuclear accidents can be potentially useful to the movement by laying bare policies and features of the power structure that are normally hidden; for example, the power of the oil companies in the Santa Barbara oil spill (Molotch 1970). Critical encounters involve face-to-face interaction between authorities and other movement actors focusing attention on movement issues. Strategic initiatives are events created by deliberate actions taken by supporters or opponents to advance movement or counter-movement goals. The staged events that are characteristic of Greenpeace campaigns are examples of this, as is the publication of polemical books such as Paul Ehrlich's *The Population Bomb* (1968), Jeremy Rifkin's *Beyond Beef* (1992) and Bill McKibben's *The End of Nature* (2006). Finally, policy outcomes are official responses to collective action by a movement or counter-movement – critical junctures at which movements are forced to renegotiate their strategies, tactics and goals as a result of changes in the political environment. The decision by the Roosevelt administration in 1914 to begin construction of the Hetch Hetchy Dam in Yosemite National Park in order to provide water for a pipeline to San Francisco was such a decision, in that it destroyed any possibility of a further alliance between the resource conservationists as represented by Gifford Pinchot and the preservationists led by John Muir.

Box 3.1 'Airpocalypse'

In early January, 2013, a vast cloud of heavy air pollution descended over China's capital city, Beijing, lingering there for most of the month, and again in February. Labelled by Internet users as airpocalypse or airmageddon, the toxic invader 'was worthy of its namesake' (Kalman 2013). Airline flights were cancelled, roads closed and the hospitals filled up with children suffering from respiratory problems. Thousands of Chinese and expatriate business executives fled the city. Tourism in February plummeted 37 per cent over the same month in 2012 (Watt 2013). The concentration of carbon particles reached a level 30–40 times higher than that deemed safe by the World Health Organization (WHO). Experts' best guess was that a sudden cold snap had triggered a spike in energy production at the region's 200-coal fired plants ('The East is grey' 2013), followed by a temperature inversion that trapped emissions under a blanket of warm air.

Following the 'airpocalypse', residents of Beijing and surrounding regions – the pollution cloud stretched 1,100 miles south – took to Weibo, China's version of Twitter, demanding that the government take action (Berg 2013). Microblogs logged 2.5 million posts on 'smog' in the month of January ('The East is grey' 2013). In stark contrast to previous episodes, 'when urban pollution was rarely acknowledged, let alone reported' (Lim 2013), the emergency received headline treatment on local and national media outlets. In an unprecedented step, the national government began releasing hourly pollution readings for 74 Chinese cities. It was, *The Economist*, observed, 'an environmental turning point' comparable to the Cuyahoga river in Cleveland, Ohio catching fire in 1969, or the outbreak of deadly mercury poisoning traceable to a spill at a plastics factory around Minamata Bay in Japan in the 1970s ('The East is grey' 2013: 18).

Staggenborg's discussion is directed primarily towards the issues of social movement mobilization and strategies, but her typology of events is relevant to the presentation of environmental claims insofar as environmental organizations often represent the primary claims-makers at this stage of the construction of environmental problems.

Of course, not all critical events are guaranteed to generate a high profile problem. According to Enloe (1975: 21), an event provokes an environmental issue when it (1) stimulates media attention; (2) involves some arm of the government; (3) demands governmental decision; (4) is not written off by the public as a freak, one-time occurrence and (5) relates to the personal interests of a significant number of citizens. These criteria are partly a function of the incident itself but also depend on the successful exploitation of the event by environmental promoters.

In presenting environmental claims, movement leaders engage in what Snow *et al.* (1986) refer to as the process of 'frame alignment'; i.e. environmental groups tap into and manipulate existing public concerns and perceptions in order to broaden their appeal. Greenpeace primarily chooses topics and organizes campaigns in areas that can lend themselves to the widest public resonance (Eyerman & Jamison 1989: 112) while avoiding those that are more divisive. In a similar fashion, environmental movement opponents attempt to appeal to a wider public by linking new technologies or programmes to popular issues and causes. Thus the biotechnology industry has tried to foster a public image of an incremental and benign technology that is useful in promoting economic development (Plein 1991). Commanding attention is not, however, sufficient to get a new issue on the agenda for public debate (Solesbury 1976: 387). Rather, emergent environmental problems must be legitimated in multiple arenas – the media, government, science and the public.

One way to achieve this legitimacy is through the use of the rhetorical tactics and strategies cited by Best (1987) and Ibarra and Kitsuse (1993). Rather than follow a chronological order, as Best suggests, environmental rhetoric has become increasingly polarized. Ecofeminists, deep ecologists and other purveyors of what Dryzek (2005) calls 'green radicalism' have tended to adopt a 'rhetoric of rectitude', which justifies consideration of environmental problems on strictly moral grounds. By contrast, environmental pragmatists, who advocate sundry versions of the sustainable development paradigm, tend towards a 'rhetoric of rationality'. Green business, for example, is based on the premise that environmentalism can be both socially useful and profitable.

This cleavage can be illustrated with reference to the loss of tropical rainforests in Brazil, Malaysia and Indonesia. Pragmatists argue that the loss of these rainforests is a serious problem because it leads to the extinction of rare indigenous insects, plants and animals that are invaluable to pharmaceutical companies as sources of new wonder drugs. Environmental purists, on the other hand, base their claims on a rhetoric that stresses the inherent spiritual value of these endangered habitats. In the same way, biodiversity concerns and biological prospecting clash when contemplating the future of the deep oceans.

Environmental claims can also be legitimated when their sponsors are acknowledged as authoritative sources of information. Hansen (1993b) has demonstrated that Greenpeace has achieved this kind of sustained success as a claims-maker in a number of ways: by acting as a conduit for the dissemination of new scientific developments between the research community and the media; by becoming a 'shorthand signifier' for everything environmental – environmental caring, green lifestyles, environmentally conscious attitudes – and by producing knowledge and information which can be used strategically in public arena debates (Eyerman & Jamison 1989).

It is sometimes possible to pinpoint an event that constitutes the turning point for an environmental problem and when it breaks through into the zone of legitimacy. With regard to global warming, this occurred at US Senate hearings in 1988 when Dr. James Hansen made the claim that he was 99 per cent sure that the warming of the 1980s was not due to chance but rather to global warming. In the case of ozone depletion, the key event was a 1988 NASA/NOAA report providing hard evidence for the first time implicating CFCs (chlorofluorocarbons) in ozone layer depletion. With pulp mill dioxins, it was the 1987 release of the '5 Mill Study' showing that traces of this toxic chemical had been detected in various household paper products and the subsequent front-page story in the *New York Times* that launched this problem in the United States, and, later, in Canada (Harrison & Hoberg 1991). It's a fair guess that the toxic smog that settled on Beijing in January, 2013 could signal an 'environmental turning point' in China (Box 3.1).

Yet scientific findings and testimony by themselves are not always sufficient to push an environmental problem past the break point of legitimacy. In the case of global warming, Dr. Hansen's earlier Senate testimony in 1986, where he predicted that significant global warming might be felt within five to 15 years, did not attract comparable coverage or concern. This only occurred two years later when there had been a significant shift in media practices and public attention (Ungar 1992: 492). Similarly, Molina and Rowland's 1974 publication in the journal *Nature* of their theory that CFCs were destroying the ozone layer at first only brought limited coverage in the California press. It was only later on when the issue became linked to claims that other gases from aerosol cans, notably vinyl chloride, were linked to skin cancer, that their data were given wide attention and media legitimacy (Mazur & Lee 1993: 686).

Contesting environmental claims

Even if an emergent environmental claim manages to transcend the threshold of legitimacy, this does not automatically ensure that an ameliorative action will be taken. As Gould *et al.* (1993: 229) have noted, one can interpret environmental protection history from the position that environmental movements have been far more successful in getting listed on the broad political agenda than in getting their policies within this agenda, especially where these policies might require the reallocation of resources away from large-scale capital interests and state bureaucratic actors.

Solesbury (1976: 392–5) has cited a number of factors that can contribute to an issue being lost at the point of decision or action. Major external constraints such as the onset of a national economic crisis may lead to a problem being postponed and then altogether abandoned. A problem may be transformed into a less threatening political issue. Opponents within government bureaucracies may use a number of tactics – postponing discussion, referring an item back for further research or amendment – which ensure that a problem will not immediately be acted upon. For example, as I write this, draft rules for energy efficiency standards proposed by Barrack Obama in his 2013 Presidential State of the Union address had been stalled for over two years at the White House Office of Information and Regulatory Affairs (OIRA) awaiting mandatory review, a fate that some think may also befall his promised initiative to combat global climate change (Verchick 2013).

As a consequence, invoking action on an environmental claim requires an ongoing contestation by claims-makers seeking to effect legal and political change. While scientific support and media attention continue to constitute an important part of the claim package, the problem is principally contested within the arena of politics. Contesting an environmental problem within the political policy stream is a fine art, given the cross pressures which legislators face.

Environmental entrepreneurs must skillfully guide their proposals through a log jam of vested and often conflicting political interest groups, each of which is capable of stalling or sinking the proposals. As Walker has noted,

> Public [environmental] policies seldom result from a rational process in which problems are precisely identified and then carefully matched with optimal solutions. Most policies emerge haltingly and piecemeal from a complicated series of bargains and compromises that reflect the biases, goals and enhancement needs of established agencies, professional communities and ambitious political entrepreneurs.
>
> (1981: 90)

Kingdon (1984) observes that policy proposals that survive in this political jungle usually satisfy several basic criteria.

First, legislators must be convinced that a proposal is technically feasible; that is, if enacted, the idea will work. This may not prove to be the case in hindsight; for example, the Endangered Species Act in the United States has worked out much less perfectly in its implementation than on paper. Nevertheless, a proposal must at least initially appear to be scientifically sound and politically administrable.

Second, a proposal that survives in the political community must be compatible with the values of policy-makers. Since most bureaucrats and politicians do not hold ecocentric views, solutions that reflect the 'New Ecological Paradigm' are not likely to get very far unless there is a generally perceived crisis. Instead, environmental solutions that, on the surface, appear to be neutral stand a better chance of being accepted than those that seem ideologically tinged. Furthermore, problems that are framed in utilitarian terms often go further than those that aren't. This means that arguments made with financial expediency in mind – figures and statistics translated into 'bottom-line' dollars (pounds/euros/yen) – are more likely to resonate than those that are presented solely on the basis of moral justifications (Hunt *et al.* 1994: 200–1).

Environmental policy is by no means a perfectly predictable and consistent enterprise. For example, Milton (1991) has suggested that the British government routinely adopts a contradictory approach to the environment. On domestic pollution issues it adopts a rigid, hierarchical position that retards change. This was evident in the British response to the acid rain problem. By contrast, on international environmental problems such as global warming, the UK has adopted a more entrepreneurial approach. On wildlife and conservation issues an approach that constitutes a mixture of the hierarchical and the entrepreneurial is favoured. Sometimes, an issue will rise in the policy agenda for totally unexpected reasons. This occurred with the greenhouse effect, which initially achieved the stamp of seriousness not in terms of a long-range threat to the world climate but in relation to what was basically a side issue: the environmental implications of the large-scale deployment of the supersonic transport airplane (SST) in the early 1970s (Hart & Victor 1993: 663–4).

In short, successfully contesting an environmental claim in the political arena requires a unique blend of knowledge, timing and luck. This process is often event-driven with a disaster such as the Three Mile Island nuclear accident opening up 'political windows' (Kingdon 1984: 213) that would otherwise remain closed. This is not to say that agenda-setting and legislative action are totally random, but the process is highly contingent upon a number of internal and external factors, many of which are not linked to the obvious merits of the case.

At the same time, there may also be a contest for 'ownership' of an environmental problem. This can be particularly rancorous where one of the contesting parties is drawn from the ranks of those directly victimized by a problem. There are many examples of this in the social problems field ranging from 'deviance liberation movements' such as the American prostitutes' rights campaign (Jenness 1993; Weitzer 1991) to victims' rights groups; for example, those formed by breast cancer patients. This is less common with environmental problems, which generally have a more diffused impact. One significant example, however, is the dispute over the issue of who owns 'biodiversity' both as a resource and as an environmental problem (see Chapter 8). This struggle pits a coalition of small farmers, ecological activists and others in Southern countries against the conservation establishment: biologists, bureaucrats from non-governmental organizations and government ministries dealing with trade and environmental issues.

Hawkins (1993) has identified three ideal-type paradigms that occupy the increasingly contested discourse over global environmental futures. The prevailing 'global managerialist paradigm' advocates the detection and solution of problems in the globalized commons by an existing configuration of nation states and international organizations buttressed by scientific experts and professional environmentalists within international NGOs (non-governmental organizations). This approach downplays local perceptions and definitions of problems, and on occasion may even blame poor people in Southern nations for causing environmental degradation. The 'redistributive development paradigm' recognizes the need for greater equity in matters pertaining to development and the environment. It proposes that such inequities can be redressed through a number of innovative measures such as the Green Fund within the World Bank or debt-for-nature swaps. The 'new international sustainability order paradigm' calls for a fundamental restructuring of the world order such that poorer nations claim a more direct voice in establishing a balance between economic and social sustainability.

Hawkins depicts the construction of international environmentalism as reflecting an ongoing struggle among supporters of these three paradigms. The dispute over the ownership of biodiversity is one manifestation of this; the conflict over global climate change is another. Even the language used in defining this contested ground is itself socially constructed. For example, countries of the North have adopted a globalized language to describe the situation in Southern nations in which 'our' environmental problems (climate change, ozone depletion) are caused by 'their' development problems (forest loss, overpopulation), a situation solvable only by embracing 'sustainable development' strategies (Redclift & Woodgate 1994: 64–5). At present, the first two paradigms still predominate but the new international sustainability order paradigm appears to be making some significant inroads.

It is important to recognize that this three-stage model of environmental issue construction needs to be adjusted so as to take into account local contexts and conditions. Martens (2007: 68) has correctly pointed out that in China the state remains more or less dominant in all of the arenas in which environmental frames are negotiated and legitimized. For example, the Chinese media are largely state owned and closely monitored. Accordingly, journalists remain careful not to take serious political risks and the frames that enter the public debate through media publications remain heavily influenced by official state policies.[5]

Box 3.2 *Necessary factors for the successful construction of an environmental problem*

- Scientific authority for and validation of claims
- Existence of 'popularizers' who can bridge environmentalism and science
- Media attention in which the problem is 'framed' as novel and important
- Dramatization of the problem in symbolic and visual terms
- Economic incentives for taking positive action
- Recruitment of an institutional sponsor who can ensure both legitimacy and continuity

Audiences for environmental claims

In addition to the skill of claims-makers and the severity of the underlying condition itself, the success of a putative environmental claim may also be tied to the magnitude of audiences that are mobilized around that claim. A groundswell of audience support not only marks the rising of a problem but it can also constitute a valuable resource in the effort to capture political attention.

For sociologists, the problem is how to reliably gauge the size and influence of audiences. As Ungar (1994: 298) has pointed out, the potential for environmental claims-makers to use public opinion as a resource is paradoxically both enhanced and limited by present polling procedures. Public polling today rarely maps support for contested positions, opting instead for broad measures of environmental concern such as the 'New Environmental Paradigm Scale' developed by Riley Dunlap and his colleagues. This produces such a vague barometer of public opinion that virtually any group on the 'pro-environmental' side can claim to represent it but, at the same time, it makes it difficult to gauge specific reactions to specific issues. Alternatively, one can look to other indicators of public support – recycling behaviour, green consumerism, participation in environmental events and mobilizations – but these too are imperfect measures of opinion.

Nevertheless, the tide of public opinion can sweep a claim upwards onto the policy agenda, sometimes in a dramatic fashion. In the 'Alar' controversy in the United States, public fears about toxins translated into a short-run consumer boycott of apples, even though the risk-supporting data were later found to be less reliable than was originally thought. Similarly, public concern about 'Mad Cow Disease' in Britain was sufficiently grave for governments to act in a precautionary manner not always so evident in the case of other potential risks. Of course, not all environmental claims succeed in raising the red flag for concerned audiences. Some claims are perceived as being too extreme, too misanthropic or too complex. Others run up against powerful counter-claims. Some fail because the requisite preventive or mitigative response mandates too great a lifestyle sacrifice.

In considering why some environmental claims capture the public eye and others do not, it may be helpful to look to the field of advertising research. In a large-scale comparative study in the 1990s, which examined the attitudes of 30,000 consumers in 21 countries, the New York advertising agency Young & Rubicam came up with a marketing model, the 'Brand Asset Valuator', which isolates four key factors that predict how well a specific product will do in the market-place: uniqueness, relevance, stature and familiarity (Scotland 1994).

In the case of environmental claims, uniqueness or *distinctiveness* refers to the extent to which the public perceives a problem as separate from others of a similar nature. For example, acid rain claims-makers were successful in distinguishing this condition from the more inclusive category of air pollution. Rhetorical strategies are important here in creating distinctive labels for emerging problems as well as devising symbolic codes that can be attached to a claim in order to confer a distinctive identity.

Relevance refers to the degree to which a particular environmental problem matters to the ordinary citizen. This is not always easy to demonstrate, even when the problem is occurring in people's own backyards. It is especially difficult where global environmental problems have their origins far away in distant parts of the world. Thus, extended drought conditions in the poor African nations are of little relevance in the southwestern United States, yet regional water shortages that require that local citizens stop watering their lawns and filling their swimming pools are quite meaningful.

Stature denotes how highly a consumer thinks and feels about a particular brand. In the case of the environment, this refers to the attitudes of the public towards the place or people or species under threat. It is no accident that the wildlife protection movement first mobilized in the nineteenth century over the danger posed to our much-loved songbirds by hunters and by the millinery trade. Similarly, national parks and monuments – Yellowstone Park in the United States, Lake District in Britain, Great Barrier Reef in Australia – have considerable symbolic stature, which comes into play if these places are imperiled. By contrast, low-income African American and Hispanic communities in the US South that face serious threats from toxic polluters have long been accorded low stature, especially by middle-class audiences.

Finally, *familiarity* refers to how well known a particular problem is to an audience. The media play an especially important role here in educating us about environments, species or places that may have been beyond our realm of personal experience. For example, in 1992 it was announced that scientists in Central Vietnam had discovered the *sao la*, a goat-like mammal previously unknown to the outside world. Almost overnight, the *sao la* became a media superstar as a result of a media frenzy whipped up by scientists, environmentalists and the press.[6] Celebrated on the pages of *National Geographic* and *People* magazines, it became 'the zoological equivalent of finding a new planet' (Shenon 1994). In some cases, environmental activists may undertake collective action in order to familiarize audiences with a claim. Thus, the clear-cutting practices in the old growth forests in British Columbia, Canada, became widely known in Europe and America, in part because of the extensive media coverage of protests by environmental activists on the logging roads and on the steps of the provincial legislature. Rather than enhancing the stature of a claim, however, familiarity may ultimately produce issue fatigue on the part of the general public, especially if new developments are not forthcoming. This is the case even if a problem is both distinctive and relevant. Indeed, audiences have an inherent sense of fair play that dictates that activities such as unrelenting 'polluter bashing' are unacceptable, even if the criticism is well deserved.

Successful environmental claims, then, must possess elements of vitality and stature that ensure that they will not perish in a sea of disinterest or irrelevance.

Necessary factors for the successful construction of an environmental problem

It is possible to identify six factors that are necessary for the successful construction of an environmental problem (see Box 3.2). These are as follows.

First, an environmental problem must have scientific authority for and validation of its claims. Science may well be an 'unreliable friend' to the environmental movement as Yearley (1992) has suggested, but nevertheless it is virtually impossible for an environmental condition to be successfully transformed into a problem without a confirming body of data that comes from the physical or life sciences. This is especially so with the newer global environmental problems, whose very existence hinges on a novel scientific construction (see the discussion of biodiversity loss in Chapter 8).

Second, it is crucial to have one or more scientific 'popularizers' who can transform what would otherwise remain a fascinating but esoteric piece of research into a proactive environmental claim. In some cases (Edward Wilson, Paul Ehrlich, Barry Commoner), the popularizers may themselves be employed as scientists; in others (e.g. Bill McKibben, Jonathan Porritt, Jeremy Rifkin) they are activist authors whose knowledge of science comes secondhand. Whatever their background, these popularizers assume the role of entrepreneurs, reframing and packaging claims so that they appeal to editors, journalists, political leaders and other opinion-makers.

Third, a prospective environmental problem must receive media attention in which the relevant claim is 'framed' as both real and important. This has been the case for most contemporary problems: ozone depletion, biodiversity loss, rainforest destruction, global warming. By contrast, other significant environmental problems fail to make the public agenda because they are not considered especially newsworthy. For example, in many Canadian cities lack of treatment of urban sewage is endemic but this has received scant coverage compared to other pollution problems. As the executive director of the Sierra Legal Defense Fund once pointed out, a volume of sewage equivalent to 32 oil-tankers the size of the *Exxon Valdez* is dumped each day into local rivers or bays, yet it is done out of the sight of the public with virtually no attention from the media (Westell 1994).

Fourth, a potential environmental problem must be dramatized in highly symbolic and visual terms. Ozone depletion was not a candidate for widespread public concern until the decline in concentration was graphically depicted as a hole over the Antarctic. The wanton practices of the major forestry companies only became a matter for international outrage when Greenpeace and other environmental groups began to exhibit dramatic photographs of the 'clear-cuts' on Vancouver Island while labelling the area the 'Brazil of the North'. Images such as this provide a kind of cognitive short cut compressing a complex argument into one that is easily comprehensible and ethically stimulating.

Fifth, there must be visible economic incentives for taking action on an environmental problem. For example, the case for acting boldly to stop biodiversity loss was levered on the argument that the tropical rainforests contained an untapped wealth of pharmaceuticals that would disappear for ever if nothing were to be done. At the same time, environmental claims that carry positive, economic incentives for one group may also involve costs for others, thus provoking sharp opposition.

Finally, for a prospective environmental problem to be fully and successfully contested, there should be an institutional sponsor who can ensure both legitimacy and continuity. This is especially important once a problem has made the policy agenda and legislation is sought. Internationally, this can be seen in the important role played by agencies and associated with the United Nations. Thus, the earthquake engineers who launched the International Decade for Natural Disaster Reduction (IDNDR) in 1990 were not having much luck getting the word out until they decided to approach the UN General Assembly to request that the Decade be operated under its auspices (Hannigan 2012a: 65).

Conclusion

Environmental concerns do not normally attract public notice and secure solutions because they are instantly and universally perceived as being serious and troublesome. Rather, they must be identified, researched, promoted and argued in the form of 'claims'. The task of 'constructing' an environmental problem is carried out by a formidable cast of 'claims-makers': scientists and medical professionals, social movement activists, journalists, policy entrepreneurs and 'victims'. While overlapping, this usually happens in three stages. In the first stage, *assembling*, the problem is discovered, named and documented, most often within the forum of science. In the second stage, *presenting* the problem is the key task. This requires commanding attention and legitimating the claim, most notably in the mass media. In the third stage, *contesting*, the key activities required are invoking action, mobilizing support and defending ownership. This occurs mostly within the forum of politics. Not all environmental problems, of course, are solvable through legislative action but it is difficult to secure effective environmental change bypassing the legal and political systems. In successfully constructing an environmental problem, we can look to six necessary factors: scientific authority for and validation of claims, existence of 'popularizes', especially in science, media attention, dramatization in symbolic and visual language, the creation of economic incentives for acting on a claim and recruitment of an institutional sponsor who can ensure both legitimacy and continuity.

Notes

1   Ibarra and Kitsuse (1993) also outline a set of 'counter-rhetorical strategies', which are meant to block claimants' attempts to construct a problem and/or demand action.
2   Another typology, Ronald Mitchell's 'processes of international agenda setting for environmental problems', generally parallels my model, but it didn't appear until well

after the first two editions of *Environmental Sociology* were already in print. Mitchell (2009) proposes that issue emergence progresses through three sequential processes: identification and recognition of the problem and its causes, diffusion of that knowledge and mobilization of concern and prioritization of the issue on the international agenda. A fourth process, 'framing the issue', cuts across and links the other three.

3   This was suggested at the public hearings on the proposed Alberta-Pacific bleached Kraft pulp mill in northern Alberta by Cindy Giday from the Northwest Territories, who was the lone native (and female) on the Alpac EIA Review Board.

4   Total membership of the 12 or so major national environmental organizations in the US increased from about four million in 1981 to roughly seven million in 1988 (Bramble & Parker 1992: 317).

5   The relationship between the media and the state in China is complex and evolving. According to Mol and Carter (2007: 14) environmental issues and campaigns are increasingly reported in a more open fashion in the Chinese media. However, they qualify this statement, noting that this freedom has 'caused greater uncertainty among journalists and media decision-makers about what is real and what is not allowed'. Generally speaking, reporting that touches on minority issues and national security and economic issues, or is directly critical of Chinese political leaders, is not permitted, unless cleared from above.

6   Unfortunately, the *sao la* soon faced extinction as collectors from around the world attempted to obtain one, even reputedly offering a bounty of up to US$1 million (Shenon 1994).

Further reading

Irwin, A. (2001) *Sociology and the Environment*, Cambridge: Polity Press.

Pettenger, M.E. (ed.) (2013) *The Social Construction of Climate Change: Power, Knowledge, Norms, Discourses*, Aldershot: Ashgate.

Yearley, S. (1992) *The Green Case: A Sociology of Environmental Issues, Arguments and Politics*, London: Routledge.

4   Environmental discourse

In recent decades, *discourse analysis* has emerged as an increasingly influential method for analyzing the production, reception and strategic deployment of environmental texts, images and ideas. Although closely identified with social constructionism, nonetheless discourse analysis has been practiced with good results by subscribers to other 'schools' of environmental theory and research, most notably by critical theorists, political ecologists and international policy analysts.

Susan Sontag, the famed essayist and literary critic, once quipped that discourse has such a prominent place in our culture simply because 'one cannot think without metaphors'(Sontag 1991: 91; cited in Davidson & Gismondi 2011: 22). Metaphors, of course, constitute only one of the components of discourse. Nevertheless, they act as a kind of powerful narrative shorthand in depicting and simplifying complex environmental phenomena. Carolan (2006) illustrates this by studying the use of metaphorical terms in three environmental science journals (*Society & Natural Resources, Conservation Biology, Ecology*). Metaphors such as alien species, habitat disturbance, ecosystem recovery and even climate change, he found, contribute to a *naturalizing discourse* wherein strong beliefs are embedded about what nature *should* look like in its pristine state, devoid of interference from humans. Carolan refers to the increasing presence of 'value creep', which he says is what happens 'when values become increasingly embedded within environmental terms that are presented as objective, value-free, and scientific' (Carolan 2006: 923).

By way of definition, discourse is an interrelated set of story lines that interprets the world around us and becomes deeply embedded in societal institutions, agendas and knowledge claims. These story lines have a triple mission: to create meaning and validate action, to mobilize action and to define alternatives (Gelcich *et al.* 2005: 379).

Discourse is the most general category of linguistic production and subsumes a number of other tactics and devices including narrative (the writing and telling of stories) and rhetoric. Some rhetoricians have drawn the ire of critical realists by insisting that we can *only* conceive of nature and the environment through the discursive language that we have developed to talk about the natural world. However, a more temperate view is that the environment as it exists in the public policy sphere is the product of discourse about nature established by scientific disciplines such as biology and ecology, government agencies, bestselling books such as Rachel Carson's *Silent Spring* and the messages disseminated by environment activists (Herndl & Brown 1996: 3).

Some analysts have been criticized for overstating the importance of discourse in environmental politics and policy making. For example, Hajer (1995: 6) insists that interests are constituted *primarily* through discourse, thereby excluding other institutional practices and institutions. Environmental politics is especially vulnerable to discursive manipulation (Hajer & Versteeg 2005). One might even say that national states largely maintain their legitimacy not by establishing sound environmental policies, but rather by invoking carefully crafted discourse that resonates with the mainstream public (Davidson & Gismondi 2011: 26; Davidson & MacKendrick 2004). Lidskog (2001) takes issue with this, arguing that discourses are by no means the only determinant of social life. The discursive dimension, he points out, is 'only one of many that are relevant to sociological analysis' and, therefore, it is problematic to claim discourse analysis, as useful as it can be, effectively constitutes a 'general approach to environmental sociology' (p. 124).

Studying environmental discourse

Within environmental studies, discourse has been visualized in a variety of ways, ranging from a 'story line' that provides a signpost for action within institutional practices (Hajer 1995), to a social movement 'frame' that enables the practices of environmental movement organizations (Brulle 2000) to an environmental 'rhetoric' constructed around words, images, concepts and practices (Myerson & Rydin 1996).

One comprehensive attempt to organize the analysis of environmental discourse comes from Herndl and Brown (1996). Their 'rhetorical model for environmental discourse' takes the shape of three circles, each of which is located at the tip of a triangle. At the top of the triangle is what they call *regulatory discourse* – disseminated by powerful institutions that make decisions and set environmental policy. Nature here is treated as a resource. At bottom right of the triangle is the *scientific discourse* where nature is regarded as an object of knowledge constructed via the scientific method. Policy makers routinely ground their decisions here, relying in particular on technical data and expert testimony. Finally, directly opposite this on the bottom left is *poetic discourse* that is based on narratives of nature that emphasize its beauty, spirituality and emotional power. Nature writing is one example of this. Herndl and Brown stress that these three powerful environmental discourses are not mutually exclusive or pure, however, and often end up being mixed together. In such cases, what we best look for are 'dominant tendencies' (p. 12).

Another effort directed at the classification of environmental discourses is Brulle's (2000) typology of discursive frames adopted by the US environmental movement. Drawing on the environmental philosophy literature and on his detailed reading of the history of American environmentalism, Brulle came up with nine distinct discourses: manifest destiny (exploitation and development of natural resources gives the environment value that it otherwise lacks); wildlife management (the scientific management of ecosystems can ensure stable populations of wildlife remain available for leisure pursuits such as sport hunting); conservation (natural resources should be technically managed from a utilitarian perspective); preservation (wilderness and wildlife must be protected from human incursion because they have inherent spiritual and aesthetic value); reform environmentalism (ecosystems must be protected for human health reasons); deep ecology (the diversity of life on earth must be maintained because it has intrinsic value); environmental justice (ecological problems reflect and are the product of fundamental social inequalities); ecofeminism (ecosystem abuse mirrors male domination and insensitivity to nature's rhythms) and ecotheology (humans have an obligation to preserve and protect nature since it is divinely created). Brulle argues that this multiplicity of discourses have resulted in the fragmentation of the American environmental movement, preventing it from speaking with a single, unified voice to a wide national audience. Adherents of each discursive frame talk past each other 'in a process of mutual incomprehension and suspicion' (p. 273). As does Schnaiberg and his entourage (see Chapter 2), Brulle concludes that there can be no meaningful environmental action without real structural change. This is unlikely to occur as long as discourses about the environment continue to block or mask the social origins of ecological degradation and proclaim a coherent vision of the common environmental good.

A third work that explicitly utilizes the typological method is John Dryzek's (2005) book *The Politics of the Earth: Environmental Discourses*. Here, Dryzek identifies four main discourses: survivalism, environmental problem solving, sustainability and green radicalism. He organizes these along two dimensions: prosaic vs. imaginative and reformist vs. radical. Prosaic dimensions are those that require action but do not point to a new kind of society, while imaginative departures from the long-dominant discourse of industrialism seek to dissolve old dilemmas and refine the relationship between the economic and the environmental. Each of these can be either reformist (adjusting the status quo) or radical (requiring wholesale transformation of the political–economic structure). According to this typology, problem solving is prosaic/reformist; survivalism is prosaic/radical; sustainability is imaginative/reformist and green radicalism is imaginative/radical. Each of these four types is, in turn, subdivided. Problem solving, for example, comes in three forms: administrative rationalism, democratic pragmatism and economic rationalism, while sustainability has two flavours: sustainable development and ecological modernization. For the most part, this typological exercise is helpful, although at an empirical level it requires some discriminating judgement calls on the part of the analyst as to what is imaginative and radical and what isn't.

There are many other books and articles, of course, that discuss environmental discourses but do not propose typologies. Two of the best known of these deal with specific 'policy' discourses: Maarten Hajer's (1995) detailed analysis of the social construction of an ecological modernization discourse on acid rain in Britain and the Netherlands in the 1980s and 1990s, and Karen Litfin's (1994) account of changing international discourse about global ozone layer depletion in the 1980s. Killingsworth and Palmer's (1996) article on 'apocalyptic' environmental discourse spans the period from the publication of Rachel Carson's *Silent Spring* in the 1960s up to more recent debates over global warming and climate change.

More recently, Craig Calhoun (2004) has identified a discourse of 'complex emergencies'. A discourse of emergencies, Calhoun tells us, is central to international affairs and now is the primary term for referring to a range of catastrophes, conflicts and settings for human suffering. As such, it serves to organize a cluster of gradually developing, predictable and enduring events and interactions into a 'crisis' that is sudden, unpredictable and short-term. This constitutes, Calhoun says, 'a discursive formation that shapes both our awareness of the world and decisions about possible interventions into social problems' (p. 376).

In this chapter I offer up my own typology (Table 4.1). As with Brulle's nine discursive frames, the three environmental discourses presented here (Arcadian, Ecological, Environmental Justice) follow a rough chronological order, as each first rose to prominence at a different stage in the history of the environmental movement. In common with Herndl and Brown's model, a distinguishing characteristic is the predominant 'motive' or 'justification' for the environmental action.[1]

I begin with an account of the emergence in the early twentieth century of *Arcadian discourse*, which, in Herndl and Brown's terms, would be described as 'poetic discourse'. In contrast to the other three discourses, Arcadian discourse peaked before the advent of the modern environmental movement in the early 1970s. Even so, the nature protection movement of the late nineteenth and early twentieth centuries acted as 'the advance guard of environmentalism' (Killingsworth & Palmer 1996: 43, note 4) and thus significantly shaped contemporary perceptions and views.

A typology of environmentalist discourse

Arcadian discourse

Writing in the 'Common Ground' column of the British newspaper *The Guardian*, Robert Macfarlane (2005) offers a thoughtful elegy for the wilder landscapes of the British Isles. Every day, he observes, millions of people find themselves deepened

Table 4.1 Typology of key environmental discourses in the twentieth century

	Discourse		
	Arcadian	*Ecosystem*	*Justice*
Rationale for defence of environment	Nature has priceless aesthetic and spiritual value	Human interference in biotic communities upsets the balance of nature	All citizens have a basic right to live and work in a healthy environment
Iconic books	*My First Summer in the Sierra*	*Silent Spring* *A Sand County Almanac*	*Dumping in Dixie*
Primary nesting place	Back to nature movement	Biological science	Black churches
Key alliance/fusion	Preservationists and conservationists	Ecology and ethics	Civil rights and grassroots environmentalism

Source: Author.

and dignified by their encounters with these landscapes. Macfarlane knows this because he has come upon testimonies in the form of graffiti, memorabilia and even poems tacked up on walls. While distancing himself from those who regard wild landscapes as 'a site for the exercise of middle-class nature sentiment', nevertheless he urges his readers to rediscover the tradition of nature writing that slipped from view a half-century ago. This is vital because such landscapes are rapidly disappearing in what novelist John Fowles has called the era of 'the plastic garden, the steel city, the chemical countryside'. In lamenting the near abolition of remoteness and celebrating its pleasures, Macfarlane is evoking what has come to be called 'Arcadian discourse'.

Van Koppen (1998: 74–5) assigns three defining features to Arcadian discourse: externality, iconization and complementarity. *Externality* means that Arcadian nature is constructed as something separate from human society, or at least removed from everyday life in the city. *Iconization* suggests that the image of nature in the Arcadian tradition is modeled on stereotyped visual images that become embedded in cultural memory. In earlier centuries these were to be found in Dutch and English landscape painting, but today they are associated more with photos of primordial wilderness settings such as the Amazon rainforest. Finally, the Arcadian tradition is best understood within the context of its *complementarity*; it stands in counterpoint to the urban industrial society and to the social and environmental ills attached to it.

In *Landscape and Memory*, Simon Schama (1995) notes that there have always been two kinds of Arcadia: one infused by lightness and bucolic leisure, the other darker and a place of 'primitive panic' (p. 517). While it is tempting to see these two landscapes of the urban imagination aligned against one another, Schama maintains that over the course of human history they have, in fact, been 'mutually sustaining' (p. 525). Much the same point has been made by the environmental historian William Cronon (1996), who describes the pivotal concept of the 'wilderness' as having its origins in two broad sources: the 'doctrine of the sublime' as conveyed in the work of nineteenth-century Romantic artists and writers such as Wordsworth, Emerson and Thoreau; and the more recent notion of the 'frontier' as proclaimed by the American historian Frederick Jackson Turner. The convergence of these two discursive elements accelerated and coalesced in the *Back to Nature Movement* in the late nineteenth and early twentieth centuries, thereby 'clothing the wilderness in a coat of moral values and cultural symbols that has lasted right up to the present day' (Hannigan 2002: 315).

Wilderness as a discursive invention: the 'Back to Nature' movement in early twentieth-century America

As Europe and America became increasingly urbanized at the end of the nineteenth century, views towards nature began to undergo a major transformation. In particular, the concept of 'wild nature' as a threat to human settlement that had long predominated gave way to a new, intensely romantic depiction in which the wilderness experience was celebrated.

The traditional image of nature and its inhabitants as frightening is reflected in much of our past and present mythical literature. For example, wolves play a central role in fairy-tales such as *Little Red Riding Hood* and *Peter and the Wolf* and more recently in the Disney film version of *Beauty and the Beast*, making the woods seem a dangerous place for children to wander alone. Similarly, readers are advised to keep out of the forest at night to avoid spectres such as the Headless Horseman in *The Legend of Sleepy Hollow*. The contemporary equivalent of this is the Hollywood horror film (*Cabin Fever, Cabin in the Woods, Resident Evil 2*) wherein a group of teenage party animals heads for an isolated cabin, only to meet a gruesome death delivered by the resident homicidal maniac, pack of zombies or supernatural forces. Civilization is depicted here as the conversion of untamed natural landscapes into a more refined pastoral setting. Note, for example, Tolkien's contrast in *Lord of the Rings* between the gentle, civilized, rolling vistas of the hobbit settlements and the wilder, darker world of the forest and mountains inhabited by walking trees, orcs and other threatening creatures.

This unfavourable attitude towards untamed nature was especially heightened during the settlement of the American frontier:

> Wild country was the enemy. The pioneer saw as his mission the destruction of the wilderness. Protecting it for its scenic and recreational value was the last thing frontiersmen desired. The problem was too much raw nature rather than too little. Wild land had to be battled as a physical obstacle to confront

and even to survive. The country had to be 'cleared' of trees. Indians had to be 'removed'; wild animals had to be exterminated. Natural pride arose from transforming wilderness into civilization, not preserving it for public enjoyment.

(Nash 1977: 15–16)

By the last part of the nineteenth century, however, a revised view of unmodified nature had emerged. Rather than a threat, wilderness was now seen as a precious resource. This view was especially strong in the United States where the frontier was on the verge of 'closing'. In the Eastern portions of the country, natural landscapes were rapidly disappearing as urban growth proceeded. Urban expansion, in turn, seemed to produce a surfeit of noise, pollution, overcrowding and social problems. In this context, unspoiled natural settings took on a special meaning; the stress of city living created a rising tide of nostalgia among the urban middle classes for the joys of country life and outdoor living.

Schmidtt (1990) studied the Back to Nature movement that flourished in the United States from the turn of the century to shortly after World War One. This movement or 'wilderness cult' (Nash 1967) encompassed a wide range of activities including summer camps, wilderness novels, country clubs, wildlife photography, dude ranches, landscaped public parks and the Boy Scouts. While it was not the only factor, this nature-loving sentiment played a significant role in the creation of the natural parks system. In the process, wild nature was transformed from a nuisance to a sacred value. As Charles Adams wrote in the *Naturalist's Guide to the Americas* (1926), the wilderness, like the forests, was once a great hindrance to our civilization; now, it must be maintained at great expense because society cannot do without it (Schmitt 1990: 174).

It is quite clear from Schmitt's and other accounts that this back to nature movement and the 'Arcadian myth' that it promulgated was socially constructed. While its supporters had mixed motives, they generally shared a belief that a return to nature represented a more wholesome set of values from those to be found in the increasingly unhealthy and corrupt environment of the city. Claims about the virtue of nature were made in each of the major institutions of the day. Leading American educators such as G. Stanley Hall, Francis Parker and Clifton Hodge actively encouraged nature study in the schools as a means of counteracting urban vices and building character. Religious educators, convinced that Americans could best find Christian values out of doors, promoted a form of pastoral Christianity in a number of ways: nature sermons, outdoor church camps, sponsorship of Scout troops and so on. Nature journalists published a steady stream of nature lore, essays, outdoor pictures and literary tales, for example, Jack London's *Call of the Wild*, celebrating the lure of wild nature. The case for wilderness preservation was taken on by a clutch of new conservation organizations such as the Sierra Club (1892) and the Wilderness Society (1935). This preservationist sentiment was especially strong among bird-watchers and ornithologists who participated in a series of crusades for over 50 years in both Britain and the United States to protect wild birds from hunters, poachers, feather merchants and other enemies (Doughty 1975).

The 'Back to Nature' movement gained a number of prominent political and institutional sponsors. None was more important than Teddy Roosevelt who, as governor of New York and then as president, became a staunch advocate of wildlife preservation. Another key supporter was David Starr Jordan, the first president of Stanford University, whose voice in support of nature study gave the movement credibility and prestige (Lutts 1990: 28). A number of important figures in the movement were based in public institutions: the American Museum of Natural History, the Smithsonian, the Carnegie Institution and the New York Zoological Society, to which they were able to bring considerable resources – money, publicity, prestige – to their preservationist and other activities on behalf of nature.

It was from these institutions also that many of the key popularizers of nature protection originated. For example, the movement to save the redwoods contained several leading scientific popularizers of the day: New York lawyer and author Madison Grant;[2] Edward E. Ayer, head of the Chicago Museum of Indian History; Gilbert Grosvenor, founder of the National Geographic Society and Fairfield Osborn a key figure in the growth of the New York Museum of Natural History (Schrepfer 1983: 41). Perhaps the highest profile popularizer (next to Teddy Roosevelt) was William T. Hornaday, for many years director of the New York Zoological Society (Bronx Zoo), who was a major force in lobbying Congress to tighten hunting regulations. Hornaday, a tireless self-promoter, wrote several widely distributed volumes on wildlife preservation as well as numerous articles in the *New York Times* and other popular publications. John Muir, the founder of the Sierra Club, was a charismatic promoter of wilderness protection who waged the country's first nationwide environmental publicity campaign during the Hetch-Hetchy controversy.

Popularizers such as Hornaday and Muir, as well as other claims-makers within the broad back to nature movement, were highly successful in garnering media attention. In this age of magazines, nature study essays and outdoor adventures were frequently featured in *Outlook*, *The Atlantic Monthly*, *Forest and Stream*, *Saturday Review*, *National Geographic* and other popular periodicals. In addition, various campaigns initiated their own publications, some of which developed a large readership. For example, the bird preservation movement spawned *Bird Lore*, the *Audubon Magazine* and other similar periodicals. *Boy's Life*, a monthly picture magazine that capitalized on the growing popularity of scouting, sold a cumulative total of 41 million issues from 1916 to 1937 (Schmitt 1990: 111). One environmental campaign, the crusade to save Niagara Falls (1906 to 1910) was waged primarily in the pages of American popular magazines, notably the *Ladies Home Journal*; it resulted in over 6,500 letters written in support of the preservation of the Falls (Cylke 1993: 22).

The back to nature movement drew upon a deep wellspring of existing cultural sentiments and in turn created a number of readily identifiable symbols and icons: the horse, Black Beauty,[3] the California redwood trees, the Grand Canyon, Old Faithful geyser in Yellowstone National Park and even Smokey the Bear. Some of these were real, others fictional creations. Nonetheless, as Schmitt (1990: 175) notes, 'those who dealt in symbols and myths found the wilderness a major force in shaping American character'.

Ecosystem discourse

A second major discourse that has powerfully shaped how we regard nature and the environment is that centering on the notions of 'ecology' and the 'ecosystem'. Referring to Herndl and Brown's (1996) terminology, we could say that the dominant tendency here is 'scientific discourse', although, as we will see, in the 1970s this fused with a normative strain within the emerging environmental movement.

Ecology has a long history prior to its ascendancy as the cornerstone of the contemporary environmental movement. Worster (1977: xiv) observes that while the term ecology did not appear until the latter part of the nineteenth century, and it took almost another 100 years for it to become a household word, the idea of ecology is much older than the name. Ernest Haeckel, the leading German disciple of Darwin, is generally credited with officially coining the term in 1866 under the name *oecologie*, by which he meant 'the science of relations between organisms and their environments'.

The full development of plant ecology owed more, however, to plant geographers, most notably the Danish scholar Eugenius Warming, who published his classic work *Plantsomfund* (The Oecology of Plants) in 1895. Warming's central thesis was that plants and animals in natural settings such as a heath or a hardwood forest form one linked and interwoven community in which change at one point will bring in its wake far-reaching changes at other points (Worster 1977: 199). This is, of course, a central message in the contemporary ecological outlook.

By the 1920s biological ecology was prospering. Two of the major figures in its development were Frederic Clements and Arthur Tansley, who developed a distinctively twentieth-century branch of biology called 'dynamic plant ecology' or 'ecosystem ecology'. Clements, a Nebraska scientist who spent most of his career as a research associate at the Carnegie Institution of Washington, is best known for his study of ecological succession. He visualized the process of succession as going from an embryonic ecological community to a more or less permanent 'climax community' that was in equilibrium with its physical environment. Once formed, it was difficult for potential plant invaders to compete successfully with established species within this climax community. However, a number of external environmental factors – forest fires, logging, erosion – might damage or destroy the climax and force succession to begin again (Hagen 1992: 27).

Tansley,[4] a British plant ecologist, is generally credited with coining the term 'ecosystem' in the mid 1930s. He strongly opposed Clements' use of the word 'community' to describe the relationship of plants and animals within a certain locale, maintaining that it was misleading because it wrongly suggested the existence of a social order (Worster 1977: 301). Instead, he came up with the concept of the 'ecosystem', which he described in terms of an exchange of energy and nutrients within a natural system. Catton calls the ecosystem the most central and incisive concept in the foundation of modern ecology, especially in Tansley's original understanding of the term, which was meant to 'unify' our perceptions of nature's units (1994: 81).

McIntosh (1985) has depicted the views of Clements, Tansley and other scientific ecologists of this era as being somewhat ambivalent with regard to human society. On the one hand, there was an acknowledgement that ecology had much to contribute to the understanding of human affairs. Clements (1905: 16) observed that sociology is 'the ecology of a particular species of animal and has, in consequence, a similar close association with plant ecology'. Tansley (1939) anticipated the establishment of a worldwide ecosystem 'deriving from inter-dependence' and stated that human communities 'can only be intelligently studied in their proper environmental setting'. While it is probably an exaggeration to state that ecology is the scientific arm of the conservation movement (McIntosh 1985: 297–9), nevertheless many ecologists have individually been active in conservation causes. Tansley himself contributed towards the campaign to establish nature reserves and later (1949) served as the first chair of the British Nature Conservancy. In the 1940s, he led efforts (mostly unfulfilled) among ecologists to establish research linkages with the four British forestry societies on the grounds that post-war plans for giant new forest plantations would cause soil fertility to suffer as well as introducing an alien feature into the aesthetics of the countryside (Bocking 1993: 92–3).

Yet, at the same time, ecologists and their societies were somewhat nervous of becoming too involved in political or social issues, fearing that their scientific credibility would be damaged. Both the British and American ecological societies were reluctant to engage in overt advocacy of particular positions or in political lobbying (McIntosh 1985: 308). Any synthesis of animal and plant ecology with human ecology was discouraged by the failure of the Chicago School of Sociology in the 1920s and 1930s to adequately conceptualize the field.[5]

By the early 1970s, however, ecology had become the theoretical cornerstone of the new and rapidly diffusing concern with the environment. Ecologists increasingly began to step outside their role as scientists to become major contributors to the environmental debate. A number of new terms were added to the English language: *ecopolitics*, *ecocatastrophe* and *ecoawareness* (Worster 1977: 341).

A British magazine, *The Ecologist*, became a centre of gravity for left-leaning environmentalists under the guidance of Edward Goldsmith.

There are several key factors that explain the centrality of ecosystem ecology in the rise of environmentalism in the 1970s.

First, the language and logic of ecology was linked to rising concerns about radioactive fallout, pesticide poisoning, overpopulation, urban smog and the like to produce what appeared to be an inclusive scientific theory of environmental problems. Rubin (1994) has argued that the instrumental force in effecting this transformation was a small group of influential writers and thinkers – Rachel Carson, Barry Commoner, Paul Ehrlich, Garrett Hardin – who functioned as scientific popularisers. Carson, in her book *Silent Spring* (1962), brought the concepts of ecology, food chains, the 'web of life' and the 'balance of nature' into the popular vocabulary for the first time. Using ecology as the explanatory linchpin, she simplified a variety of problematic relationships into one 'environmental crisis' (Rubin 1994: 45). Commoner (1971) systematized Carson's observations with his four laws of ecology: 'everything is connected to everything else'; 'everything must go somewhere'; 'nature knows best'; 'there is no such thing as a free lunch'. These laws may have over-simplified ecosystem ecology but they had enormous rhetorical power. Similarly, Hardin's (1978) metaphor of the 'tragedy of the commons' found a broad audience both within the academic world and outside.

Second, the fusion of ecology and ethics first achieved in Aldo Leopold's 'land ethic' featured prominently. The land ethic was first proclaimed in his book *A Sand County Almanac*, published posthumously in 1949. It extended ethical rights to the natural world, which he regarded as a community rather than a commodity. In the 1950s, Leopold's work had a small but committed following in conservation circles but only became widely known after it was reprinted in 1968. Whereas the ecosystem had previously been largely a theoretical construct, albeit a dynamic one, now it was inculcated with moral significance. Human interference in biotic communities not only had particular effects, for example, forcing a new round of succession as Clements had suggested, but it was also defined as the wrong thing to do. This insight became especially significant with the rise of 'deep ecology' in the 1980s.

Finally, as Macdonald (1991: 89) has observed, by co-opting scientific ecology the environmental movement added considerably to its strength for two reasons. First, despite the fact that ecosystem ecology was considered to be a somewhat 'soft' discipline within the natural sciences, it nevertheless allowed environmentalists to claim the authority of science for their campaigns. Second, because of its holistic perspective, ecology attracted a variety of 'seekers' such as devotees of expanded consciousness, Zen and organic food who might otherwise have had little interest in green causes. Combined with scientific ecologists, these newcomers created a potent political mix. In recent years, this alliance has been at best an uneasy one but in the early 1970s it brought the idea of an 'ecological threat' into the pervasive currents of alternative popular culture where journalists constantly troll in their search for the emergence of new trends.

Ecology, then, was transformed from a scientific model for understanding plant and animal communities to a kind of 'organizational weapon',[6] which could be used to systematize, expand and morally reinvigorate the environmental. In the process, it acquired a new texture: more political, more universal and more subversive (Sears 1964; Shepard 1969). While some scientific ecologists reacted negatively to this reconstitution of the concept, others supported it, arguing that the 'environmental crisis' demanded a new sense of social activism on the part of biological researchers. The latter became influential claims-makers, presenting a politicized vision in which the boundaries between nature and society were deliberately blurred.

Kinchy and Kleinman (2003) have identified the existence of two competing discursive tendencies within contemporary scientific ecology – purity and utility. On the one hand, it is argued that ecology is a value-free, objective science with legitimate claim to expertise. At the same time, many academic ecologists have explicitly aimed to demonstrate the discipline's usefulness in the policy-making arena. The Ecological Society of America (ESA), the primary professional scientific society for ecologists in the United States, has attempted to deal with these pressures by undertaking various programmes and initiatives designed to reap the benefits of public engagement while asserting the value neutrality of the discipline. For example, in 1979, having concluded the credibility of ecology was being sullied by non-experts claiming to be ecologists, the ESA established a voluntary certification programme designed to enable ecologists to participate in public debates over environmental issues while protecting the autonomy and unique expertise of ecology as a whole (pp. 882–3).

Most recently, the meaning of ecology has once again undergone yet another reconstruction. Grassroots activists such as those found in the Chipko Movement in India and the Greenbelt Movement in Kenya have proposed a new alternative ecological perspective in which insight into ecosystem interrelationships is achieved by means of folk knowledge rather than scientific observation. Indigenous wisdom of this type is embedded in practices that preserve the environment in the long run. Alas, local ecological knowledge has been suppressed by the juggernaut of global economic development, which forces the poor off their ancestral lands and deprives them of the opportunities to follow sustainable practices (Clapp & Dauvergne 2005: 109).

This alternative knowledge system provides citizens' movements with the tools for the reconstructing science. In this context, ecology becomes part mythology, part popular science; a rallying point for opposition to the kind of environmental diplomacy that predominated at the Rio Conference. As such, it represents a fresh social interpretation of a 130-year-old concept even if it is one that might be unrecognizable to Haeckel, Warming and other pioneers of scientific ecology.

Environmental justice discourse

In the 1980s, a new discourse emerged in the United States that differed dramatically from prevailing ones in its interpretation of environmental problems and priorities. Environmental justice (EJ) has become increasingly used as part of the language of environmental campaigning, as a description of a field of academic research, as a policy principle and as a name given to a political movement (Walker 2012: 1).[7] EJ thought, Dorceta Taylor (2000: 508, 566) observes, has 'altered the nature of environmental discourse and poses a challenge to the hegemony of the NEP'.

Environmental justice lays out a set of claims concerning toxic contamination in terms of the 'civil rights' of those affected rather than in terms of the 'rights of nature' (Nash 1989). Capek (1993) identifies four major components of this environmental justice frame: the right to obtain information about one's situation; the right to a serious hearing when contamination claims are raised; the right to compensation from those who have polluted a particular neighbourhood and the right of democratic participation in deciding the future of the contaminated community. Each of these components represents a specific claim that has been rhetorically formatted in the language of 'entitlement' (Ibarra & Kitsuse 1993).

Whereas the concept of ecology was utilized in the 1970s to join together rising concerns about toxic pollution with an ethical concern for nature, environmentalism in the 1980s and 1990s underwent another transformation in which the central discourse is *environmental justice*. This shift occurred primarily at the grassroots level both domestically and in the Southern hemisphere. While some key figures in this movement wanted to throw off the environmental label entirely,[8] others framed their claims to justice and equity within the context of an environmental movement. Environmental justice activists have not totally abandoned the legacy of the previous two decades: Commoner's industrial-ecological critique, for instance, has been one theoretical referent for this alternative explanation of the roots of the environmental crisis. At the same time, concerns about resource conservation, wilderness preservation and pollution abatement are downplayed in favour of issues such as the uneven distribution of resources and development and the safety of minority workers.

In the introduction to a special issue of the journal *Qualitative Sociology* on the topic of social equity and environmental activism, Schnaiberg (1993: 203) laments the failure of environmental sociology to consider social inequality. As early as 1973, Schnaiberg claims, he was writing about the political necessity of incorporating elements of social justice into any proposal for environmental action but this message fell on deaf ears. This may reflect shortcomings in the field, as Schnaiberg suggests, but it is also a reflection of what was going on within the environmental movement itself.

In both the United States and Britain, the mainstream environmental movement was (and to some extent continues to be) dominated by a relatively narrow set of concerns; for example, rural planning and wildlife preservation. These are said to reflect the white, middle-class membership of the main environmental organizations.

In the United States a number of health-related environmental inequities were exposed during the 1960s and 1970s but they rarely made it into the larger movement agenda. Gottlieb (1993) highlights the differential treatment given to three issues during this period: pesticide poisoning, the toxicity of lead and uranium hazards.

For migrant farm workers in California the explosion of pesticide use through the 1950s and 1960s created a number of health-related problems. In its successful campaigns for farm workers' rights during these years, the United Farm Workers (UFW) under the leadership of the charismatic Cesar Chavez aggressively initiated legal suits to obtain information about the chemical ingredients in sprays, as well as campaigning to ban specific pesticides including DDT and to have pesticide-related health and safety language incorporated into UFW-grower contracts.[9] Aside from some limited assistance from the Environmental Defense Fund, the mainstream environmental movement generally avoided the question of human exposure to pesticides, focusing primarily on the impact of pesticides on wildlife, as indeed did Rachel Carson.

During the 1960s, childhood exposure to lead paint became a significant local issue in a number of inner city communities in the United States. By 1970, Gottlieb (1993: 247) notes, dozens of inner city-based community groups and coalitions were organizing to address lead paint issues, primarily in East Coast cities such as Rochester, Washington, New York and Baltimore. Aided by New Left-inspired groups such as the Medical Committee for Human Rights, the lead paint movement achieved significant visibility both locally and nationally. At this point, however, the emphasis shifted to lead levels in the air, especially as a result of the emission of leaded gas (petrol). Mainstream environmental groups such as the Natural Resources Defense Council and the Environmental Defense Fund that had previously avoided involvement in this issue put a priority on reforming the Clean Air Act, eventually forcing a ban on the sale of leaded gas. The lead paint issue did not re-emerge until the late 1980s and by then the primary claims-makers were from alternative environmental groups within the social justice movement.

Starting in the 1950s, uranium poisoning began to affect thousands of transient prospectors, mine and mill workers and residents of communities living downwind of the uranium mines. This 'radioactive colonization' (Churchill & La Duke 1985) was concentrated among native American workers in New Mexico and Arizona. For example, a 1979 spill of radioactive tailings into the Rio Puerto in northern New Mexico contaminated significant stretches of Navajo Indian lands. As Gottlieb (1993: 251) observes, the Rio Puerto spill occurred just weeks after the Three Mile Island accident, yet it received limited attention from policy-makers and mainstream environmentalists. Indeed, during the 1950s conservation groups ignored uranium issues altogether because they were perceived as occurring far from the scenic wilderness sites celebrated as part of an Arcadian discourse. During the following decade, environmental groups were primarily concerned with nuclear power as an alternative energy choice, although the anti-nuclear movement had begun to gather strength. Only in the 1990s did some groups accord the toxic threat to Indian lands a higher priority.

In each of these three cases, mainstream environmental groups focused on separate though oftentimes parallel concerns, defining them in 'environmental' rather than 'social justice' terms (Gottlieb 1993: 253). In constructionist language, they established 'ownership' of the problems on behalf of a primarily upper-middle-class or elite Anglo constituency. On a more general level, they focused mainly on regulation or containment rather than seeking to subvert the social order in order to bring about a form of social reconstruction that would benefit 'have not' constituencies (Hofrichter 1993: 7).

In what proved to be somewhat of an anomaly, the Conservation Foundation, an organization whose brief focused largely on research and education, convened a conference in Woodstock, Illinois, in November, 1972 to explore the themes of race, social justice and environmental quality. At this gathering, urban planner Peter Marcuse, son of the famed social philosopher Herbert Marcuse, presciently warned participants that divorcing equity and social justice concerns from the environmental agenda threatened to create a permanent rupture (Gottlieb 1993: 253). It would be another decade and a half, however, until this rupture started to reach the public eye and the environmental justice discourse started to attract attention.

Emergence and growth

In the United States, the environmental justice movement didn't emerge until the early 1980s. As Bullard (1990: 35) notes, this newfound activism 'did not materialize out of thin air nor was it an overnight phenomenon'. Rather, it was the result of a growing hostility by urban blacks in the United States to the siting of toxic landfills, garbage incinerators and the like in neighbourhoods or communities with predominantly minority populations. In the 1970s this discontent was confined largely to the local context but within the decade it spread to a wider theatre as the struggle for environmental equity was presented as a fight against *environmental racism*.[10]

There are several key milestones in the emergence and growth of the environmental justice movement during this period.

In 1987, the Commission for Racial Justice of the United Church of Christ (UCC) released an influential report, *Toxic Wastes and Race in the United States*, which documented and quantified the prevalence of environmental racism. The UCC report firmly established the 'grounds' for this claim, by setting out the magnitude of the problem in numerical terms. Among its findings was the revelation that three out of five black Americans live in communities with uncontrolled toxic waste sites. Furthermore, blacks were heavily over-represented in those metropolitan areas with the greatest number of such sites: Memphis, Tennessee; St Louis, Missouri; Houston, Texas; Cleveland, Ohio and Chicago, Illinois, with over 100 toxic waste sites each. Hispanics, Asian Americans and native peoples were similarly over-represented in high-risk communities. This confirmed a study conducted four years earlier by the US General Accounting Office that reported that three of the four largest commercial landfills in the South were located in communities of colour.[11]

Also crucial in establishing the dimensions of environmental racism was the research of sociologist Robert Bullard. In 1979, Bullard, who taught at the predominantly black Texas Southern University in Houston, was invited by his wife, a lawyer, to conduct a study on the spatial location of all of the municipal landfills in that city in order to provide data for a class action lawsuit that she was arguing. Bullard confirmed that toxic waste facilities, not only in Houston but elsewhere in America, are most likely to be found in African American and Hispanic urban communities. In a series of journal articles beginning in 1983 (many co-authored with Beverly Wright) and in his book, *Dumping in Dixie*, Bullard documented these environmental disparities and the mobilization of the environmental equity movement. Even more than was the case for the UCC report, Bullard's work established the 'warrants' for this problem, arguing that action was justified in order to reclaim for minorities 'the basic right of all Americans – the right to live and work in a healthy environment' (1990: 43). Bullard has since become a key leadership figure as indicated in 1992, when he was chosen by the Clinton administration to participate in the Presidential transition process as a representative of the environmental justice movement (Miller 1993: 132).

In January 1990, Bunyan Bryant and Paul Mohai,[12] professors in the School of Natural Resources at the University of Michigan, organized the University of Michigan Conference on Race and the Incidence of Environmental Hazards with papers from 12 scholar–activists. Among the follow-up strategy of the Michigan Conference were a series of meetings in Washington with key government officials including William Reilly, Administrator of the EPA, and Congressman John Lewis (Bryant & Mohai 1992). One of the important outcomes of these meetings was Reilly's decision to create an internal EPA working group to investigate the evidence and draft a set of proposals to address environmental inequalities (Mohai 2008: 24).

Third, under the sponsorship[13] of the Commission for Racial Justice, the First National People of Color Environmental Leadership Summit was held in October 1991 in Washington, DC. At this gathering, three strands of environmental equity were identified (Lee 1992): *procedural equity* (governing rules, regulations and evaluation criteria to be applied uniformly), *geographic equity* (some neighbourhoods, communities and regions are disproportionately burdened by hazardous waste) and *social equity* (race, class and other cultural factors must be recognized in environmental decision-making). Delegates to the Summit ratified a document, *Principles of Environmental Justice*, which sets out the ideological framework of the emerging environmental justice movement. Taylor (2000: 537–42) sorts these principles into six thematic components: ecological principles: justice and environmental rights; autonomy/self determination; corporate–community relations; policy, politics and economic processes; and social movement building. While grounded in the ecocentric principles espoused by the pioneers of the environmental movement (John Muir, George Marsh, Aldo Leopold), the principles also argue that people have a right to clean air, land, water and food and the right to work in a clean and safe environment. The document affirms the rights of people of colour to determine their own political, economic and cultural futures. It strongly opposes military occupation and exploitation of their land and calls for their participation as equal partners in the policy arena. To ensure environmental justice, the *Principles* call for strong social movement building, both nationally and internationally. Gottlieb (1993: 269) credits the summit with advancing the environmental justice movement past a 'critical threshold in definition' both by ratifying a common set of principles and by identifying a new kind of environmental politics of inclusion.

Organizationally, the movement has been held together in a number of decentralized, loosely linked networks of umbrella groups, newsletters and conferences (Higgins 1993: 292) rather than the top-down, professionalized configuration typical of mainstream environmentalism. This has its roots in the formation in the early 1980s of several nationally based 'anti-toxics' groups – the Citizens' Clearinghouse for Hazardous Wastes and the National Toxics Campaign. In the mid-1990s, the emphasis shifted somewhat from national to regional grassroots networks, as epitomized by the Southwest Network for Environmental and Economic Justice.

While the social construction of discourse and strategic framing were crucial in communicating the environmental justice message to supporters, these were not by themselves sufficient, Taylor (2000: 563–4) notes, to mobilize a strong base of supporters. Instead, the EJM embraced several key recruitment strategies. Rather than try to build movement networks from scratch, organizers tapped into lines of preexisting social relationships and networks, drawing from networks of people with past histories of social and political connections. In particular, they targeted people with strong institutional ties to churches, labour unions, universities, community organizations, federal agencies, legal institutions, grant givers and mainstream environmental organizations. Having observed the growth throughout the decade of the 1990s of Federal Government offices, programmes and initiatives devoted to pursuing environmental equity, the latter (mainstream environmental groups) finally started collaborating with communities of colour and EJOs, 'slowly diversifying their staff and memberships, covering environmental justice issues in their magazines, launching environmental justice programmes, and undertaking a variety of environmental justice initiatives' (Taylor 2000: 559). Pellow (2000) gives a good example of this changing relationship. During the 1970s and 1980s, many environmentalists in the Chicago area endorsed the growing waste-to-energy incinerator industry as a way of converting trash into a useful form. By the 1990s, however, they reversed their support for incineration, in large part because they observed activists in communities of colour and working-class neighbourhoods engaged in struggles to resist industry's efforts to site these facilities near their homes.

One mainstream environmental group that has signed on to the environmental justice agenda is the Sierra Club. In 1993, the Sierra Club adopted its first environmental justice policy, stating that 'to achieve our mission of environmental protection and a sustainable future for the planet, we must attain social justice and human rights at home and around the globe' ('Joining together for justice' 2004). Since then, the Club has undertaken a number of initiatives: providing organizational assistance to over 250 low-income neighbourhoods and communities in the US; hiring full time environmental justice organizers in Detroit, Washington, DC, the southwest and central Appalachia; awarding grants to help local groups to put together and lead 'toxic tours', create community gardens and undertake public education programmes about the environmental damage caused by factory farms in the South; helping block or shut down polluting mines, incinerators and sewage treatment plants; and collaborating with Amnesty International to defend activists under threat from the state for speaking out on environmental issues.

In the early 1990s, the environmental justice movement expanded its charter to incorporate the exploitation of people in Southern countries. Much of the interaction between grassroots activists from the United States and their counterparts in the South has taken place in the context of the United Nations; for example, at the Rio Summit and its preparatory meetings. Environmental justice activists from the US also participated in a 1992 meeting hosted by the Third World Network in Malaysia that focused on toxic waste. These networking activities with Third World activists have moved environmental justice leaders back on the path to a renewed ecological awareness. Vernice Miller, a co-founder of the group West Harlem Environmental Action, described environmental justice as a 'global movement that seeks to preserve and protect global ecosystems' (1993: 134).

Note, however, that tensions have arisen since the early 1990s between US-based environmental justice activists and their compatriots in the global South. David Pellow (2007) notes the relative absence of American EJ groups in transnational environmental justice coalitions. He attributes this partly to their insistence on working mainly through networks of people of colour, while resisting joining forces with allies of different ethnicities, races and nationalities around the world. By contrast, activists in the global South 'have made the pragmatic decision to join forces with allies across borders to increase their leverage at home and elevate the visibility of their struggles beyond their domestic national spheres' (Pellow 2007: 234). On their part, EJ activists in other nations have not always been eager to link up with American groups. In South Africa, for example, activists have strategically not wanted to be seen to be simply following a US-based discourse and model of campaigning, insofar as the United States is regarded as being deeply implicated in patterns of economic and environmental exploitation around the world, and in the causes of global-scale problems such as climate change (Kalan & Peck 2005; Walker 2012: 36).

Anand (2004: 15) has drawn a parallel between the themes of the American environmental justice movement and international environmental politics between North and South. In particular, he identifies inequities in the international arena relating to both procedural and distributive justice that are similar to the national politics of environmental justice in the United States. Just as the environmental justice movement in the United States represents a backlash against the failure of government to address gross inequities in exposure to toxic dumps and other health hazards, there has been 'tremendous opposition to many international global agreements and efforts because they do not adequately reflect the interests of countries of the South' (p. 15). The Biodiversity Convention is a leading example of this (see Chapter 8). Furthermore, the power imbalances inherent in the global economic system lead to situations where lower income residents in the nations of the South are differentially impacted environmentally. Not only is this a case of differential exposure to industrial effluents and other pollutants, but it has also meant inequities in access to basic natural resources such as fuel and drinking water.

As in Herndl and Brown's (1996) rhetorical model of environmental discourse, the three environmental discourses discussed above should not be treated as being either static or mutually exclusive; rather, they engage one another in dialectical fashion. For example, in Southern nations, environmental activists have successfully merged elements of the ecosystem and social justice discourses. Creating 'new imaginaries' helps energize local struggles and draw in a more diverse set of allies by 'giving the demands of subordinate groups a new claim to universality' (Evans 1992: 8–9). Over time, Dryzek (2005: 20) tells us, environmental discourses 'develop, crystallize, bifurcate and dissolve'; and, sometimes, they return in a different wrapping.

Accordingly, Arcadian discourse, whose zenith in America passed nearly a century ago, has re-emerged in recent decades in the form of a romantic and spiritual celebration of the Amazon rainforest. Thus, Slater (2002: 101) describes current images and accounts of the Amazon such as those that are used to attract eco-tourists as having a 'dual nature'. They intertwine 'virgin' and 'virago' – traditional narratives of a 'lush, dark, exciting jungle' that is harsh and untamed and con-temporary narratives of a fragile rainforest composing an intricate web of flora and fauna. This latter image, in turn, connects with ecological discourse that portrays the rainforest as 'both a storehouse of valuable commodities and a key to global environmental health' (p. 138). A final discursive layer imbues this exotic, biotic paradise with an extra layer of moral urgency by drawing on a justice frame to publicize the plight of rubber tappers (notably Chico Mendes who was murdered in 1989), Indian tribes (Kayapó, Yanomani, Machiguenga) and other indigenous forest dwellers that face displacement and decimation at the hands of ranchers, miners and other agents of development.

Discourse, power relations and political ecology

It is difficult to talk about discourse these days without bringing in a discussion of power; mostly, this is due to the influence of the French social theorist Michel Foucault (1979, 1980) who transformed our theoretical understanding of power, as well as putting discourse analysis on the academic agenda. Most analytic perspectives in the humanities and social sciences that employ the concept of discourse 'have Foucauldian elements in terms of viewing discourses as something that defines what is meaningful and how it exercises power' (Gelcich *et al.* 2005: 379).

Foucault dismissed the previously paradigmatic notion that power necessarily resides permanently in institutions, notably the state; rather, he conceptualized power as being embedded in social relationships. As such, it is not just a weapon wielded by the ruling class but a fundamental feature of everyday human interaction. As Hindness (2001: 100) phrases it, power is manifested 'in the instruments, techniques and procedures that may be brought to bear on the actions of others'. This can range from the power of a president or prime minister to control the agenda of a national news conference, to the power of one partner in a marriage to control the choices of television programmes made by his or her spouse.

Power may be everywhere but relationships of power are rarely symmetrical and wholly democratic. Here, Foucault makes an important distinction between power and domination. The latter refers to asymmetrical relationships of power in which the subordinated party has a negligible chance of exercising his or her will. Whereas power relationships are often unstable and reversible, domination means that these relationships are less fluid and less open to negotiation.

In the case of interpersonal power relations, one individual may hold power over another due to superior physical strength, attractiveness to others, rhetorical abilities, higher income or social status or a more extensive network of political contacts (Scott 2001: 135). Where power is structured around formal institutions such as the state and the corporation, other means are required. In such cases, Foucault believed that power is exercised not so much through naked force and physical coercion as through the ability to shape the process of socialization. This is much more effective because it reduces resistance while internalizing consent. It is at this point that discourse becomes important.

Discourse provides institutions with a powerful means of incorporating individuals into relations of domination. Foucault regarded this as central to a process of social control that he labelled 'discipline'. Increasingly, he observed, this occurs under the supervision of 'experts' who are empowered through their stranglehold on scientific and technical forms of discourse (p. 92). While Foucault was primarily concerned with the exercise of discipline within total institutions such as prisons and psychiatric hospitals, his insights about the close relationship between discipline and expertise can just as easily be extended to the domains of science and environmental risk determination.

In the ongoing cultural contest in which discourse is shaped, some players possess more resources than others. In their cross-national study of talk about abortion, Ferree *et al.* (2002) coin the concept of a 'discursive opportunity structure', which they define as the 'complex playing field [that] provides advantages and disadvantages in an uneven way to the various contestants in framing contests' (p. 62). Here, they are drawing on the social movements literature, combining the framing approach of Snow and Benford with the political opportunity perspective of Charles Tilly and others. Ferrree and her colleagues are especially interested in the power exercised by large institutions such as the mass media, the judiciary, the churches, the political parties and social movement organizations and how this shapes the process of producing (shaping) abortion discourse (Monteiro 2005: 160).

In the environmental sphere, *legitimacy* is a key resource. This is not assessed solely by the competition for economic resources (Dowling & Pfeffer 1975: 124), but culturally through the use of discursive tools. Consider, for example, the longstanding conflict over forest preservation and conservation in the old growth rainforests of British Columbia between environmentalists and logging companies. Cormier and Tindall (2005) have shown that the former effectively utilized 'wood frames', key words and images such as that of the endangered giant spruce tree, to legitimize their campaign in the media domestically and abroad. By so doing, they provided moral ammunition for international actions such as consumer boycotts against pulp and paper companies. Conversely, resource companies can use discursive practices to legitimate their activities; thus, Davidson and Gismondi (2011: 170) argue that it was ideology, not economics, that ensured the development of the Alberta Tar Sands (Box 4.1).

Control over discourse production also carries with it the power to 'delimit both the actors that can legitimately engage in politics and the issues that are subject to debate' (Davidson & Frickel 2004: 478). This is nicely illustrated in a case study by Carolan and Bell (2004) of a dioxin controversy that flared in 2000–2001 in the small midwestern city of Ames, Iowa.

The conflict began with a report commissioned by the North American Commission for Environmental Cooperation by the internationally known scientist/environmentalist Barry Commoner and colleagues at the Center for the Biology of Natural Systems in New York. In his report, Commoner attributed accumulations of dioxin, a toxic chemical, in the Inuit community of Nunavut in the Canadian Arctic to 'drift' from a handful of major polluters in the United States. One of those cited was the waste incinerator attached to the municipal power plant in Ames. AQLN, a local activist group that had previously had some success in turning a local quarry into a park and water supply reservoir, organized a campaign of opposition. Commoner himself was brought to Iowa State University to deliver

---

Box 4.1 *Legitimating the Tar Sands*

The Alberta Tar Sands (in the resources sector this is known as the Alberta Oil Sands) is a mega project that has become the focus of a bitter North American energy conflict, insofar as it is the production point of the oil that would flow through the controversial Keystone XL Pipeline. In Challenging Legitimacy at the Precipice of Energy Calamity, Debbie Davidson and Mike Gismondi (2011) discuss tar sands development and politics with a view to explaining how public investment and support was secured for decades of investment, research and marketing. What they discovered is that authors of the project framed it to the provincial government and to the public within the tent of three well-accepted narratives: a Westernized worldview of frontier individualism, a utilitarian view of ecosystems and confidence in the continued progress. In so doing, they recognize that what matters most to legitimacy is not what supportive conditions are present, but rather that evidence of contradictions is notably absent (2011: 172). Freudenburg and Alario (2007) describe this as diversionary reframing.

---

a lecture on 'Globalization and the Environment'. In response, the City commissioned Robert Brown, an Iowa State engineering professor, to conduct a study as to whether the power plant should be tested. Citing scientific inadequacies in the Commoner report (lack of data directly connecting incinerator emissions with Arctic dioxin build-up; apparently incorrect information on plant use and construction), Brown recommended against in-plant testing.

Citing Foucault's observation that social relations are also relations of power, Carolan and Bell (2004: 287) stress that the city government and the university engineers effectively controlled the public debate. AQLN's 'threat to public health' frame never effectively competed with an official 'the Ames power plant is safe and a source of community pride' frame. Other than Commoner, the activists' voices were rarely heard in the local press. As relative newcomers to the city, AQLN members were not well integrated into local community networks. By contrast, 'those with access to the dominate social networks also have an avenue through which to express their opinions and have these options heard, all of which has bearing on how others perceive them as speaking the truth' (ibid.).

Discourse and political ecology

Discourse and discursive clashes play a central role in recent scholarship that follows the terrain of what has been called the *new political ecology*. This takes the form of locality-based studies of people interacting with their environments (Goldman & Schurman 2000: 568). Contemporary political ecology, Evans (2002: 8) explains, 'arose out of a dissatisfaction with traditional versions of ecological arguments, which tended to ignore the dilemmas of people whose livelihood depended on the continued exploitation of natural resources'.

Researchers following a political ecology approach have focused on environmental struggles related to North–South relations. In this context, the term 'South' is used not only in a geographic sense to refer to Asia, Africa and Latin America, but also to reflect 'the common experiences of people in these countries as a result of historically determined social and economic conditions resulting from their colonial and imperial past' (Anand 2004: 1).

Escobar (1996) has argued persuasively that capitalist development today is routinely sheathed in seemingly beneficial discourses such as sustainable development and biological conservation. This is easy to do because their meaning is, at the very least, ambiguous. The underlying purpose here, however, is always to 'capitalize nature'.

Goldman and Schurman (2000: 570) identify two ways in which political ecology scholars have usefully employed discourse analysis: (1) as a method of understanding alternative discourses on nature, the environment and environmental degradation and how these clash with dominant discourses imposed by the state, Northern environmental movements and transnational NGOs and (2) as a means of exploring and exposing the power relations embodied in national and global conservation agendas (see Chapter 8).

As we have seen, formulating and communicating ecological discourse is not restricted to those in power, although the latter have the upper hand in making their discourses dominant. There is a growing literature in the social sciences that discusses alternative discourses on nature and the environment that flow from the grassroots. These discourses are rarely unopposed, however, since they inevitably challenge the state and other claimants to local land and natural resources.

Paul Ciccantell (1999) makes the important point that the discursive struggle in a socially remote extractive periphery such as the Brazilian Amazon is usually a matter of powerful external actors imposing their 'definitions' over the objections of indigenous groups. He illustrates this with a case study of the Tucuruí dam, built on the Tocantis River in the eastern Amazon in the 1980s as a joint venture between a Japanese private and government consortium and the Brazilian government. The Tucuruí dam effectively cut off all river transport, forced the relocation of 20,000–30,000 people and transformed local ecosystems thereby causing threats to human health (e.g. malaria), local climatic changes and a change in plant and animal species.

In this and other similar cases, three principal meanings formulated by the Brazilian military government in the 1960s and 1970s prevailed: the region's rivers were an obstacle to road-building and colonization efforts; the rivers were a source of hydroelectric power for the raw materials processing industries and growing regional population centres and waterways were access routes for oceangoing ships to export raw materials at low cost. Ciccantell stresses that a 'pluralist' model of social construction, as prevails in the United States and Europe in which competing groups seek to define issues in terms that support their interests, does not normally operate here. Rather, the discursive process is dominated by regional and national economic and political elites that are able to impose their definitions, even in the face of organized public opposition (p. 296).

Another important depiction of these discursive struggles can be found in Nancy Peluso's reporting of her research in the 1990s in Java and in the western interior of Borneo, both of which are part of Indonesia.

In an earlier study, Peluso (1992) focused on the struggle over the teak forests of Java. The Indonesian state tried to assert resource control over both the forest and its indigenous inhabitants by applying the modern legal constructs of 'property rights' and 'criminality' (those who violated the former). In response, the forest dwellers developed 'a counter-discourse on what is a fair, legal and legitimate use of the forest' (Goldman & Schurman 2000: 569).

In her later research, conducted in the province of West Kalimanata, Peluso (2003) concentrates on strategies used by local people to counter official government mapping exercises that are justified on the grounds that they are an ongoing part of territorial resource management. As part of their efforts to mount 'counterclaims or reclaims to contested or appropriated resources' (p. 232), villagers engage in *countermapping*. This is a technique by which traditional land and resource claims, many of which go back to a time before the Dutch colonized Indonesia, are expressed in a contemporary format. Using locally drawn 'sketch maps' that reflect local custom or practice, the 'mappers' – local activists sometimes assisted by international consultants – develop high-tech maps that are used to make claims to government and large international NGOs.

Peluso argues that this may be practically sound, but it contributes to the emergence of a new 'hybridized discourse' in which 'common rights in long-living trees, held communally by multiple generations of heirs, are slowly being replaced with a notion that property rights in land supersede or dominate all forms of property in trees and other territorial resources' (p. 233). In engaging in counter-mapping, NGOs and others are utilizing 'state tools' and buying into state discourses of 'territorialization'. They are also opening the door to several undesirable possibilities. Once rights to resources are mapped or documented, the state gains a certain power over these resources and the people claiming them, becoming both 'a recognized arbiter and mediator of both access and rights'. Furthermore, the conditions are established for increased community conflict, especially in regions that will likely see increased migration and intermarriage in the future.

A second way that discursive struggles may arise in the countries of the South is in response to attempts by national and international NGOs (INGOs) to impose their agendas and viewpoints on indigenous peoples. In particular, this is manifested in the tendency of INGOs to selectively take fragments of localized knowledge and translate these into the 'global' discourse of science (Dumoulin 2003: 593–4). This has been very much the case with the issue of biodiversity loss (see Chapter 8) where, until relatively recently, environmentalists from abroad were committed to establishing biosphere reserves and other protected areas, usually at the expense of local people.

Using Mexico as an example, Dumoulin (2003) demonstrates how national and international ENGOS (environmental non-governmental organizations) have successfully framed the protection of 'indigenous' knowledge within the framework of a new, global-oriented approach to nature conservation. In particular, an epistemic community (see Chapter 6) composed of ethnobiological experts with similar academic backgrounds and common values (enhancement of cultural and biological diversity for the future of humankind) have effectively exerted their influence in international arenas of power. They have done so by creating a cognitive framework, disseminating information and lobbying politically. The key group here is the International Society of Ethnobiology (ISE) in concert with the leaders of the Amazon Alliance, the Forest People Programme, the World Rainforest Movement and Cultural Survival. Under the direction of its founder, Darell Posey, the ISE has been particularly successful at securing positive media coverage and in ensuring that the issue of biodiversity-related indigenous knowledge is on the international environmental agenda (Dewar 1995).

Environmental NGOs are not always the most faithful translators of Western conceptions into Western discourses. As Roué (2003) points out, they are often no closer to the socio-cosmic view of indigenous peoples than were the resource development forces that preceded them. At best, they 'enable marginal populations, deprived of political power to acquire some at least, and to enter into communication with the centre and the dominant society' (p. 620). This is complicated further by the fact that ENGOs, by their very nature, are concerned not so much with the people to which they relate as with 'the natural environment they inhabit, which is often perceived as wild and in need of protection' (p. 621). On occasion, this can lead to misunderstanding and conflict.

One particular point of difference has been over the image of local people as noble 'defenders of nature' employing their ancestral wisdom to protect non-renewable resources. In the case of indigenous people in both the North and the South this is only true to a certain extent. Some traditional inhabitants, for example the Kayapó of the Amazon, are in fact quite entrepreneurial and are willing to sell their gold and timber for the right price (Conklin & Graham 1995; Dewar 1995; Slater 2002). As Slater (2002: 150) has noted, indigenous people in rainforest areas have proven most adept at borrowing vocabularies of human rights and environmental protection for their own ends; so too have tribal leaders in parts of Canada. In response, some disenchanted conservationists have concluded that sustainable development is impossible and that rainforests can only be protected through excluding all humans, including local people who dwell there. This has sparked a renewal of conservation programmes wherein tracts of land with relatively untouched natural ecosystems are purchased with public donations and fenced off as 'nature reserves' in order to keep them 'pure'.

The lesson here, I think, is not that rainforest dwellers or other local populations in the South are necessarily 'frauds'. Neither should it be concluded that the threats posed to the environment by mining, forestry, road-building, corporate agriculture and urban sprawl are to be discounted. Rather, it should tell us that discourses that frame the situation in simplistic terms as a conflict between 'conservation' and 'exploitation' with local inhabitants assuming the role of environmental defenders' are best treated with caution.

Conclusion

One especially influential type of social construction is 'discourse', the linguistic production of narratives, storylines and rhetoric. As we saw in earlier chapters, much of the sociological writing on the environment is divided on the basis of whether it follows an 'apocalyptic' narrative or a narrative of 'ecological improvement and sustainability'. In this chapter, I offer up a typology of environmental discourses that broadly follows the history of environmentalism, especially in America. In act one, during the late nineteenth and early twentieth century, environmentalism (or conservation as it was then known) was energized by an 'Arcadian' discourse, which held that nature has spiritual value and a priceless aesthetic. When it re-appeared at mid-century, environmentalism was now directed by an 'ecosystem' discourse that fused scientific ecology with the holistic beliefs of the counterculture movement. The key rationale here for environmental action was that human interference in biotic communities upsets the balance of nature. More recently, environmentalism has taken a turn towards a 'justice' discourse. This combined civil rights warrants with grassroots environmentalism, arguing that every citizen has a basic right to live and work in a healthy environment. I conclude the chapter by directing the discussion to the importance of considering discourse and power relations jointly, especially in studying the 'political ecology' of resource conflicts in Southern nations.

Notes

1 This contrasts with Dryzek (2005: 10) who clearly spells out his intent to 'lay out the basic structure of discourses that have dominated recent environmental politics' and to 'produce something more than just an account of environmentalism'.

2 Grant is one of the more controversial figures in the early wilderness protection movement. A patrician lawyer with close links to many elite figures in business and politics including Teddy Roosevelt, he was among other things a founder of the

Save the Redwoods League, the New York Zoological Society and the Boone and Crocket Club. At the same time, he has been called by historian John Higham (1963) 'intellectually the most important nativist in recent American history'. Grant's book, *The Passing of the Great Race* (1921), was for a while a popular exposition on the principles of eugenics although it was less successful in subsequent printings. Grant's concern with the subject of eugenics and racial exclusion was shared by a number of other leading wilderness protectionists of the day including William Hornaday, Fairfield Osborn and Vernon Kellogg.

3   Anna Sewell's book *Black Beauty* was originally published in England in 1877 where it sold more than 90,000 copies. It was re-published by the American Human Education Society in 1890. While designed to increase support for the animal welfare movement, the book also helped to establish a climate for the wider support of wildlife conservation (Lutts 1990: 22–3).

4   Tansley was eclectic in his interests and friendships, having among other things helped the social philosopher Herbert Spencer revise his *Principles of Biology* and pursued an interest in psychoanalysis by studying briefly under Freud and writing a popular book on Freudian psychology (Hagen 1992: 80). He was also an entrepreneurial scientific leader who played an instrumental role in establishing the British Ecological Society in 1913 and serving for 20 years as editor of the Society's *Journal of Ecology*.

5   In the 1930s, due largely to the efforts of Charles Adams, director of the New York State Museum and Paul Sears, a plant ecologist, the Ecological Society in the US made some attempt to bring social scientists and ecologists together in a common forum, notably in a joint symposium of the Society with the American Association for the Advancement of Science entitled 'On the Relation of Ecology to Human Welfare: The Human Situation'. Sadly, two of the leading theorists of the Chicago School, Ernest Burgess and Roderick McKenzie, were unable to attend, leaving August Hollingshead as the only representative of sociology (Cittadino 1993).

6   The term 'organizational weapon' was first introduced in Philip Selznick's classic (1960) study of the American Communist Party. Eyerman and Jamison (1989) borrow the concept to describe Greenpeace's use of flamboyant and sometimes illegal media-capturing actions to pressure governments and business. Organizations are weapons in such cases when they act in a manner that is considered unacceptable by the community.

7   In this book, I have situated my discussion of environmental justice in Chapter 4 on 'Discourse' rather than in Chapter 2 on 'Key Perspectives and Controversies in Environmental Sociology'. This reflects Walker's (2012: 1) observation that environmental justice is significant and worthy of attention because it brings to the discourse of contemporary political life 'an important way of bringing attention to previously neglected or overlooked patterns of inequality which can matter deeply to people's health, well-being and quality of life'.

8   In a 1992 interview, Lois Gibbs, the heroine of the Love Canal story, told environmental activist and author Robert Gottlieb: 'Calling our movement an environmental movement would inhibit our organizing and undercut our claim that we are about protecting people, not birds and bees' (Gottlieb 1993: 318).

9   In an article published posthumously, Chavez (1993: 166–7) charges that corporate growers in California effectively sidestepped many of the provisions of these contracts including those governing the use of pesticides. Chavez observes that many of these same growers were the largest financial contributors in the campaign to defeat Proposition 128 (nicknamed 'Big Green'), a 1990 ballot initiative supported by

environmental groups and the UFW, which among other things would have 'protected California's last strands of privately held redwoods and banned cancer-causing pesticides'.

10   The term 'environmental racism' was evidently coined by the Reverend Benjamin Chavis, former head of the United Church of Christ Commission on Racial Justice and later Executive Director of the National Association for the Advancement of Colored People (NAACP), a major civil rights organization in the United States (Higgins 1993: 287).

11   The impetus for this study was a request from Walter Fauntroy, a congressional representative from Washington, DC and an active participant in a struggle in Warren County, North Carolina, to stop the establishment of a toxic landfill containing PCB-laced soil (Bryant & Mohai 1992: 2).

12   Mohai (2008: 23) writes that in 1987, his University of Michigan colleague Bunyan Bryant, an African American environmental studies professor, pointed him to the recently released *Toxic Wastes and Race in the United States*. This UCC report 'had a major influence' on him.

13   It should be noted that the funding for this conference was gold-plated, including, among other sources, the Ford Foundation and the Rockefeller Family & Associates (Mayer 1992).

Further reading

Carson, R. (2002) *Silent Spring* (40th Anniversary Edition), Boston and New York: Houghton Mifflin.

Dryzek, J.S. (2005) *The Politics of the Earth: Environmental Discourses*, Oxford: Oxford University Press.

Hajer, M. (1995) *The Politics of Environmental Discourse: Ecological Modernization and the Policy Process*, Oxford: Clarendon Press.

Walker, G. (2012) *Environmental Justice: Concepts, Evidence and Politics*, London and New York: Routledge.

Online resources

Environmental Justice Resource Centre at Clark Atlanta University. Available HTTP: http://www.ejrc.cau.edu/

National Black Environmental Justice Network. Available HTTP: http://www.nbejn.org/who.html

5 Media and environmental communication

In moving environmental problems from conditions to issues to policy concerns, media visibility is crucial. Without media coverage the odds are low that an erstwhile problem will either enter into the arena of public discourse or become part of the political process. For example, it is unlikely that many of the lay public would have become aware of 'mad cow disease' or the purported dangers of genetically modified (GM) foods if it were not for media reportage (Lupton 2004: 187). Indeed, most of us depend on the media to help make sense of the bewildering daily deluge of information about environmental risks, technologies and initiatives.

While the traditional news media are important here (and are the focus of this chapter), there is also an extensive array of other media sources, from documentary television shows on nature and the environment to motion pictures to Internet websites. For example, MTV in the United States broadcasts *Trippin*, a conservation series directed at teenagers. Co-produced by film actress Cameron Diaz, the series presents endangered animals in their natural habitats. Episodes include Diaz in Tanzania and Nepal, gangsta-rapper DMX in the Yellowstone outback and professional surfer Kelly Slater on the Costa Rican coral reefs (Martel 2005). While it met with a decidedly polarized reaction, the controversial 'Cow Girl' billboard campaign launched by the activist group MAdGe (Mothers Against Genetic Engineering in Food and the Environment) instantly secured widespread public attention to the debate over the social and cultural ethics of genetic engineering in New Zealand (Box 5.1).

Indeed, the news media's role as an agent of environmental education and agenda setting is both important and complex. As Schoenfeld *et al.* (1979) have demonstrated, the daily press in the United States was initially slow in grasping the basic substance and style of environmentalism, leaving it to issue entrepreneurs in colleges and universities, government and public interest groups to mobilize concern outside of the media net. In local environmental conflicts, media claims are often viewed skeptically, refracted as they are through the prism of residents' own practical everyday experiences and knowledge (Burgess & Harrison 1993). Rather than actively sparking a response to environmental problems, the media

Box 5.1 'Cow Girl'

On 1 October, 2003, billboards started appearing across New Zealand's two largest cities, Auckland and Wellington, depicting a naked, genetically engineered woman kneeling on all fours with her four breasts hooked up to a dairy milking machine and the red letters 'GE' branded on her buttocks. MAdGE (Mothers Against Genetic Engineering in Food and the Environment), a protest organization founded by former pop music icon Alannah Currie Madge, claimed responsibility. In an accompanying press release ('Why Not Just Genetically Engineer Women for Milk?' 2013), Currie Madge explained what the billboard symbolically represented:

> New Zealanders are allowing a handful of corporate scientists and ill-informed politicians to make decisions on the ethics of GE. Our largest science company, AgResearch, is currently putting human genes into cows in the hope of creating new designer milks. The ethics of such experiments have not ever been discussed by the wider public. How far will we allow them to go? Where is the line in the sand? Do we as human beings have the right to blur the boundaries between species, especially when we do not know what the long term consequences may be. As an experiment, it transgresses the fundamental integrity of both woman and cow.

In their case study of protesting mothers and GM cows, Bloomfield and Doolin (2013) conclude that the Cow Girl billboard campaign was certainly successful in terms of grabbing the media spotlight and energizing supporters – the advertisements reportedly swelled onsite visits to MAdGE's website from 500 to 7,000 a day (Bloomfield & Doolin 2013: 515). However, by flouting conventional mores regarding the public representation of motherhood and the female body, and by violating the accepted boundaries between humans and animals, the billboard risked cancelling the intended message about the ethical dangers of genetic engineering. Indeed, the campaign uncorked an unexpected backlash, generating criticism from Reality Check, another anti-GM group and triggering a ruling by the Advertising Standards Complaint Board that the billboard had breached prevailing community standards.

often seem to be a millstone weighing down public discussion of environmental topics in a technical–bureaucratic discourse that excludes interest groups and non-official claims-makers (Corbett 1993: 82).

In this chapter I will assess the news media's conflicting role in socially constructing environmental issues and problems. Of particular concern is the extent to which the portrait of the environment presented by mainstream journalists represents (or not) a critique of the paradigm of technological progress as opposed to simply an extension of the existing corpus of disaster stories. First, however, it is necessary to briefly outline the general process through which the media manufacture news and endow issues and events with symbolic meaning.

Manufacturing news

For many years, mass communication researchers largely took for granted the existence of 'objective' facts and events that could be verified, exclusive of whether or not they were actually covered by the media. Thus, floods and hurricanes, political victories and resignations, medical miracles and foreign wars were all thought to have a certifiable existence of their own beyond the newsroom. Reporters, editors, producers and other 'newsworkers' might, on occasion, distort or selectively omit certain happenings but this did not mean that they were not real (Fishman 1980: 13).

In the 1970s, this approach gave way to a very different model, in which events become news only when transformed by the 'newswork' process and not because of their objective characteristics (Altheide 1976: 173). News is conceptualized here as a 'constructed reality' in which journalists define and redefine social meanings as part of their everyday working routine (Tuchman 1978). Newsmaking, in turn, is treated as a collaborative process in which journalists and their sources negotiate stories.

Organizational routines and constraints

While the construction of news may be influenced by cultural or political factors, it is generally seen as the result of inescapable organizational routines within the newsroom itself. Schlesinger (1978) observes that rather than being a form of 'recurring accident', news is the product of a fixed system of work whose goal is to impose a sense of order and predictability upon the chaos of multiple, often unrelated events and issues. In his observational study of BBC news, he found that the backbone of each day's newscasts was a routine agenda of predictable stories: labour negotiations, parliamentary business, activities of the Royal Family, sport scores, etc. In a similar fashion, Fishman (1980) observed that, rather than dig for information, reporters at a California daily newspaper opted instead for a diet of routine news derived from a mix of scheduled events (press conferences, courtroom trials) and pre-formulated accounts of events (arrest records, press releases); these items were crucial in helping them to meet deadlines and story quotas.

In addition to mandating that news be planned, time also acts as a constraint upon the final product itself. This has the effect of rendering news reports 'incomprehensible rather than comprehensive' (Clarke 1981: 43). In particular, action clips that fit more easily into existing formats, especially television news, are favoured over longer, more nuanced stories that deal with underlying causes and conditions.

By consistently failing to ask the question 'why', the news process 'contributes to *decontextualising* or removing an event from the context in which it occurs in order to *recontextualise* it within a news format' (Altheide1976: 179). This tendency is further encouraged by the use of *news angles* – frameworks around which a particular content is moulded in order to tell a story. The use of news angles is pervasive in journalism and plays a significant role in determining not only the 'spin' put on a story but also whether a story is suitable in the first place for broadcasting or publication.

Media constructionists have also noted the importance of news sources in shaping story content. Reporters usually stick to a short list of trusted source contacts who, on the basis of past experience, can be counted on to be both articulate and reliable. Indeed, it is not unknown for source contact lists to be passed down from one reporter to the next. Trusted sources come from various walks of life but they are usually people who function in official roles: politicians, the heads of government agencies, scientists and other experts. Even where the media solicit comment from opponents of the *status quo*, news sources are invariably drawn from the executive of major social movement organizations such as Greenpeace, the Sierra Club and Friends of the Earth.

In their study of the 1969 oil spill in Santa Barbara, California, Molotch and Lester (1975) found that powerful figures and organizations with routine access to the media (the president, federal officials, oil company representatives) were far more likely to function as news sources than were conservationists and local officials. These sources exercise considerable social and political power by providing a pre-packaged, self-serving, socially constructed interpretation of a given set of events or circumstances – an interpretation that is readily adopted by journalists who rarely have the time or the specialized knowledge needed to flesh out their own news angle (Smith 1992: 28).

Media discourse

In recent years, media constructionists have looked beyond the social organization of the newsroom and focused on the process by which journalists and other cultural entrepreneurs develop and crystallize meaning in public discourse (Gamson & Modigliani 1989: 2). This approach takes as its central concern the decoding of media texts – the visual imagery, sound and language produced in the social construction of news and forms of public communication (Gamson *et al.* 1992: 381).

The key element here is that of *media frames*, a concept adapted by several media sociologists in the late 1970s and early 1980s (Gitlin1980; Tuchman 1978) from Erving Goffman's work on small group interaction. Frames, like news angles, are organizing devices that help both the journalist and the public make sense of issues and events and thereby inject them with meaning. In short, they furnish an answer to the question, 'What is it that's going on here?' (Benford 1993: 678). When expressed over time, frames are known as *story lines* (Gamson & Wolfsfeld 1993: 118).

Even when the details of an event are not disputed, the event can be framed in a number of different ways. For example, the 1993 murder of Liverpool toddler James Bulger by two 10-year-old boys was variously framed by the press as a new low in the continuing economic and moral decline of England, the turning point in the campaign against 'video nasties' (one boy's father had reportedly rented the movie *Child's Play 3* just before the crime), a cautionary tale for harried parents with youngsters in tow and an example of the linkage between school truancy and juvenile crime. Both claims-makers and their opponents routinely compete to promote their favoured frames to journalists as well as to potential supporters. At the same time, news workers forge their own frames largely for reasons of efficiency and story suitability. Gamson and Wolfsfeld (1993) depict the interaction between movements and the media as a subtle 'contest over meaning' in which activists attempt to 'sell' their preferred images, arguments and story lines to journalists and editors who, more often than not, prefer to maintain and reproduce the dominant mainstream frames and cultural codes. In the Nicaraguan conflict of the 1980s, for example, peace activists attempted to counter the official frame that the American-sponsored Contras were waging a struggle against Communist expansion by promoting a 'human costs of war are too high' frame (Ryan 1991).

Finally, as Gamson *et al.* (1992: 384) point out, it is wrong to assume that news consumers (readers, audiences) passively accept media frames as they are; they too may decode media images in different ways, utilizing varying frameworks of interpretation (Corner & Richardson 1993).

Media discourse, therefore, takes the form of a 'symbolic contest' in which competing sponsors of different frames measure their success by gauging how well their preferred meanings and interpretations are doing in various media arenas (Gamson *et al.* 1992: 385).

The process of framing is comparable to the rhetoric of claims-making in social problem construction (see Chapter 3). Gamson and Modigliani (1989: 3–4) distinguish five *framing devices*: metaphors, exemplars (i.e. historical examples from which lessons are drawn), catchphrases, depictions and visual images, and three *reasoning devices*: roots (a causal analysis), consequences (i.e. a particular type of effect) and appeals to principle (a set of moral claims), which function as a kind of symbolic shorthand in telegraphing the core meaning of a frame.

Furthermore, they introduce the concept of *media packages*. Media packages help to organize these framing devices in cases of complex policy issues such as the use of nuclear power. In analyzing television news coverage, news magazine accounts, editorial cartoons and syndicated opinion columns on nuclear power from 1945 to the present day, Gamson and Modigliani isolate seven different interpretive packages: progress, energy independence, the devil's bargain, runaway, public accountability, not cost-effective and soft paths. As the titles suggest, each package is represented by a metaphor, catchphrase or other symbolic device (1989: 3).

Mass media and environmental news coverage

As Schoenfeld *et al.* (1979: 42–3) have demonstrated, prior to 1969, the daily press in the United States had considerable difficulty recognizing environmentalism as a topic separate from that of conservation. Conservation was a reasonably well-understood and respectable concern, having been around since the 1880s. It had a known constituency, its own legislative acts and administrative bureaux and even its own universally recognized symbol – Smokey the Bear. By contrast, the central tenet of environmentalism, i.e. that everything is connected to everything else, seemed difficult to grasp in journalistic terms. Similarly, in Britain, the preservation of the countryside, the national heritage and rare species of fauna and flora were all widely accepted as legitimate activities that cut across class lines, but few journalists readily connected them with air pollution, oil spills and other contemporary environmental problems.

During the late 1960s and early 1970s, media coverage of the environment rose dramatically and, for the first time, environmental issues were seen by journalists in both Britain and America as a major category of news (Lacey & Longman 1993; Parlour & Schatzow 1978). Newsworkers began to perceive individual difficulties such as traffic problems or pollution incidents as part of a more general problem of 'the environment' (Brookes *et al.* 1976; Lowe & Morrison 1984).

There are several key events that may be cited in order to explain this upswing in media awareness and understanding of environmentalists' claims. Schoenfeld *et al.* (1979: 43), citing Roth (1978), argue that the most effective environmental message of the century was totally inadvertent: the 1969 view from the moon of a fragile, finite 'Spaceship Earth'.[1] This provided a powerful metaphor with which to frame the environmental message. Similarly, Earth Day 1970 acted as a news 'peg' for a variety of otherwise disparate news stories on environmentally related subjects, earning extensive coverage both nationally and in many local American communities (Morrison *et al.* 1972).

After 1970, however, media coverage of the environment began to fall off (Parlour & Schatzow 1978), although it recovered briefly during the energy crisis of 1973–4. When stories appeared they were most likely to be event-related and problem-specific. In their examination of article headlines in the *Canadian Newspaper Index*,

Einseidal and Coughlan (1993: 140) observed that environmental items were located under a series of disparate and seemingly unconnected problem categories: air pollution, water pollution, waste management and wildlife conservation. Similarly, Hansen (1993a: xvi) notes the tendency of the media to define the environment 'largely in terms of anything nuclear (nuclear power, nuclear radiation, nuclear waste, nuclear weapons), in terms of pollution and in terms of conservation/protection of endangered species'. Rarely were the global aspects of environmental problems highlighted during this period. Even more unusual was the appearance of stories on environmental problems in Third World countries.

This pattern appears to have changed somewhat in the 1980s and 1990s. Einseidal and Coughlan note that towards the end of 1983 new descriptor terms began to appear in Canadian newspaper headlines; for example, 'global catastrophe', 'environmental order' and 'environmental ethics'. In contrast to the conservation focus, stories were vested with a more global character, encompassing attributes that included 'holism and interdependence and the finiteness of resources' (1993: 141). They also note an increasing urgency and seriousness in the coverage of environmental issues by the Canadian press as indicated by the appearance of a collection of 'war and dominance' metaphors: survival, defeat, battles, crusades. Topic headings were found to be more specific, covering such areas as 'eco-tourism', 'environmental law' and 'eco-feminism'. In a similar fashion, Howenstine (1987) detects a transformation in environmental reporting in major US periodicals from 1970 to 1982 towards a greater complexity of coverage. In addition, he found a shift across time to relatively fewer articles on the degradation and protection of the natural environment and more on economic and development issues.[2]

However, perception that coverage is deepening may have been overly optimistic. Lacey and Longman (1993) note the rise of a 'show business' and commercial approach to environmental issues in the British media during the 1980s and argue that the improvements in environmental reporting are only evident if a narrow definition of environmental issues is utilized. In particular, an artificial separation is created between the environment and development issues in line with a predominant editorial and political bias. For example, coverage of famine in East Africa has been big on shock tactics but short on political insight, especially in the case of drought in the Sudan, a country whose political regime is considered unacceptable by Western policy brokers. Furthermore, the reporting of such stories is cyclical and usually in step with their ascendancy on the political agenda (Anderson 1993b: 55).

Production of environmental news

To a large extent, media coverage of environmental issues is shaped by the same production constraints that govern news work in general. Earlier in this chapter, we discussed some of the most significant of these constraints: limited production periods, story lengths and limited sources. Clarke (1992) has grouped these production constraints into two general categories: short-term logistical and technological constraints, and long-term and occupational constraints that are embedded in the news process itself.

Short-term pressures of time have dictated that environmental issues and problems be framed by journalists within an event orientation. As Dunwoody and Griffin (1993: 47) point out, this event orientation limits journalistic frames in two ways: (1) it allows news sources to control the establishment of story frames; (2) it absolves journalists from attending to the bigger environmental picture. Three major types of environmental events can be identified: milestones (Earth Day, the Rio Summit); catastrophes (oil spills, nuclear accidents, toxic fires) and legal/ administrative happenings (parliamentary hearings, trials, release of environmental white papers).

The twin lures of celebrity and at milestone events can be seen at the 1992 Earth Summit at Rio de Janeiro in Brazil. Those attending included not only more than 100 heads of state, including US President George Bush, British Prime Minister John Major, German Chancellor Helmut Kohl and Cuban President Fidel Castro, but also an estimated 12,000 representatives from NGOs. Among the celebrities from the world of politics and entertainment were California governor Jerry Brown, actors Jeremy Irons and Jane Fonda and American media mogul Ted Turner. Even before the official summit began, a fundamental conflict arose between the wealthy nations of the North and the poorer countries of the South over a wide spectrum of issues. Finally, the summit was accompanied by an array of what *Time* magazine called 'sideshows galore' (Dorfman 1992): a fantasy ballet, *Forest of the Amazon*; an indigenous people's conference and a concert for the Life of Planet Earth. The symbolism of the occasion was typified by the giant Tree of Life in Rio on which were hung leaf postcards from children worldwide.

Thirteen years later, the 2005 summit of G8 leaders in Gleneagles, Scotland, was more or less appropriated by rock stars Bob Geldof and Bono and converted into a media opportunity to publicize their campaign to eradicate African debt and end global poverty. As a lead-up, ten 'Live 8' concerts were staged across four continents watched by two billion people worldwide. At the meeting itself, Geldof and Bono were accorded quasi-diplomatic status, meeting one-on-one with world leaders, even as the usual assortment of protesters, including environmentalists, were held back behind the barricades. By the time media interest shifted dramatically to the terrorist bombings in London, African poverty had received an unprecedented week in the media limelight.

Catastrophes are the bread and butter of environmental news coverage. They frequently involve injury and loss of life or the possibility of such. There are sometimes acts of tremendous courage or self-sacrifice. Human interest stories abound: the stubborn but proud homeowner who sits on the roof and refuses to evacuate as the floodwaters rise; the baby who is found alive after three days in the rubble of an earthquake-devastated neighbourhood.

According to Wilkins and Patterson (1990: 19), this event-centred reporting is characteristic not only of quick onset disasters such as tornadoes, hurricanes and blizzards but also of slow onset environmental hazards: ozone depletion, acid rain and so on. In order to fit these latter phenomena into the news agenda, journalists are required to picture them as the recent outcome of an event rather than the inevitable outcome of a series of political and societal decisions.

While event-centred coverage has the advantage of raising public awareness of otherwise ignored environmental topics, it also has a negative side. By focusing on discrete events rather than on the contexts in which they occur, the media tend to give news consumers the impression that individuals or errant corporations rather than institutional politics and social developments are responsible for these events (Smith 1992; Wilkins & Patterson 1990). This is especially applicable for environmental catastrophes. For example, in the case of the 1989 *Exxon Valdez* oil spill, the media framed the story in terms of Captain Joseph Hazelwood's alleged alcohol problems rather than deal with other potentially important news angles such as the recent history of cutbacks in maritime safety standards administered by the coast guard, or the oil industry's lack of capability in cleaning up large oil spills in settings such as Prince William Sound (Smith 1992). Cottle (1993: 122) has described this as the tendency of an item to remain 'entrapped within the narrow confines of its news format', unable to allow any background explanation or any input from outside, non-official voices.

Furthermore, stories about hazards favour mono-causal frames rather than frames involving long and complex causal networks. Thus, Spencer and Triche (1994) found that increases in toxic pollution in the drinking-water supply of New Orleans during the summer of 1988 were almost exclusively attributed to a simple natural phenomenon – a drop in water levels in the Mississippi River due to drought conditions – rather than to a combination of low water levels and a long-standing problem with discharges from chemical plants upriver from the city. They speculate that this monocausal framing occurred because newspaper personnel were reluctant to implicate several powerful institutional actors – the US Army Corps of Engineers, the state bureaucracy, the chemical industry – as contributors to this hazard event.

Cottle's comments further suggest a second feature of the news process that shapes the nature of coverage: a public access that is largely restricted to official news sources. Since few reporters themselves feel qualified to sort out the often conflicting scientific, technical and political claims involved in an environmental problem, they either avoid substantive issues altogether (Nelkin 1987) or turn to informed sources[3] who can offer a credible and easily summarized précis of what is happening.

While these 'primary definers' are depicted as coming exclusively from a hierarchy of social and political elites, Cottle (1993: 12) argues that this is not necessarily the case for environmental stories. Analyzing a sample of British television programmes from 1991 to 1992, he found that various diverse elements (i.e. scientists, diplomats, local officials and politicians, environmental pressure groups, individual citizens) collectively constituted the primary definers.[4] At the same time, Cottle indicates that this was by no means 'a situation of open and equal access' since environmental news clearly depends on a number of well-organized interests, some from the dominant elite, some from opposing groups. Several agenda-building studies from the same era concluded that environmental journalists in the United States prefer information from sources who, they perceive, have no obvious self-serving economic purpose (Curtin & Rhodenbaugh 2001). Business and industry sources were rarely cited here as representing reporters' first source of information (Dumanoski 1994), nor named as being a very credible source (Fico *et al.* 2000).

However, Anderson (1993) has questioned whether it is possible to deduce patterns of source dependence from content analysis alone. Supplementing content analysis with interviews, she found that ease of access varies over time. For example, during the late 1980s, Greenpeace and Friends of the Earth had good access to the national media in Britain, but they subsequently experienced some difficulty as the threat to the environment gave way to other issues such as the economic recession.

At various points in its recent history, environmental news coverage has also suffered because it does not fit easily into the structure of routine news production. Metropolitan daily newspapers tend to be partially organized according to fixed 'beats' – city hall, industrial (labour) relations, crime, sports, etc. Schoenfeld (1980: 458) cites one reporter as describing the classic environmental story as a 'business-medical-scientific-economic-political-social-pollution story'. This being so, editors and producers often don't know what to do with stories about the environment. It should be noted, however, that this may have a positive aspect, insomuch as individual environmental reporters are sometimes given considerably more leeway than their colleagues working on other journalistic beats because environmental issues are so often difficult for non-specialists to understand (Fletcher & Stahlbrand 1992: 183).

Smaller newspapers and broadcast newsrooms are less likely to use beats, opting instead for a general assignment system (Friedman 1984: 4). This, however, creates another set of difficulties. General assignment reporters, despite their optimism that they can quickly acquire adequate knowledge about subjects in which they have no background or training, are rarely capable of sophisticated reporting[5] such as that demanded by many environmental stories.

Based on his comprehensive analysis of news coverage of three environmental catastrophes – the 1988 Yellowstone Park forest fires, the *Exxon Valdez* oil spill and the Loma Prieta 'World Series' earthquake in 1989 – Conrad Smith, himself a former photographer and film editor, identifies three major difficulties experienced by such general reporters: (1) they did not conceptualize these major catastrophes as anything more than large-scale versions of warehouse fires or train derailments; (2) they did not have the structural freedom to go beyond the obvious stories; (3) they did not know how to find experts and evaluate their relative scientific qualifications (1992: 190).

When environmentalism first took off as a news story in 1969–70, many daily newspapers set up an environmental beat.[6] Reporters were recruited from allied beats – nature, outdoor recreation, science – or from the general assignment pool. While the volume of environmental coverage rose, the quality did not always keep pace. In particular, these rookie environmental reporters seemed to experience difficulty with both the substance and style of environmentalism (Schoenfeld *et al.* 1979). When the environment faded as an issue after 1970, many of these beats were shut down (Friedman 1983), although some of them were later re-commissioned (Hansen 1991).

A final short-term constrain on environmental reporting is the role and influence of news editors. With one eye always fixed on circulation or audience figures, editors tend to favour stories that feature controversy and conflict. As a result, thoughtfulness often gives way to sensationalism. In addition, editors are more likely to be sensitive to external pressures from corporate advertisers and other powerful supporters of the *status quo*. Reporters know this, and on occasion modify or deliberately overlook significant stories that involve environmental wrongdoing (Friedman 1983). This evidently occurred in the late 1970s in Houston, Texas, where local newspaper reporters were not willing to go against the predominant 'boomtown' mentality and report the problems surrounding a nuclear power plant and a nuclear treatment facility (Hochberg 1980).

Longer-term constraints on environmental journalism relate to historically evolved journalistic priorities, notably the requirements for news 'balance' and 'objectivity'. These dual pillars of objective journalism first arose during the nineteenth century as part of the sweeping intellectual movement towards scientific detachment and the culture-wide separation of fact from value (Gitlin 1980: 268). Despite periodic lapses, news-workers today still view objectivity and balance as the cornerstones of their profession.[7]

For environmental reporting, objectivity and balance mean that reporters often attempt to distance themselves and their readers from the environmentalist struggle to effect a shift in public consciousness, taking refuge instead in the objectivism of science (Killingsworth & Palmer 1992: 149). Journalists see themselves as a neutral and ironic voice, willing to be won over only if the scientific evidence concerning acid rain, global warming, biotechnology, etc. is sufficiently powerful and unambiguous. The major shortcoming of this approach is that few environmental reporters are sufficiently well informed to be able to effectively evaluate the 'scientific standing' (Friedman 1983: 25) of the evidence. Alternatively, reporters may turn to the traditional 'equal time' technique whereby both environmentalist claims-makers and their opponents are quoted with no attempt to resolve who is correct. In this case it becomes difficult for environmentalists to convince the public that an emerging 'issue' is in fact a 'problem'.

Boyne (2003: 35) has identified a tension between the media's dual imperative of analyzing risk and creating an appetite for its images. All too often, it is the latter that predominates. Journalists, editors and producers abandon the skeptical stance described above and embrace the role of a 'campaigner'. In such instances, the media actually come to *lead* the public agenda. On occasion, this can lead to considerable harm, especially where the scientific evidence is inflated or misconstrued. This is what appears to have happened in the case of cell phone 'scares' in Britain in the 1990s.

The ideal of objectivity also means that journalists rarely express the content of environmental stories in overtly political terms, opting instead for news frames that emphasize conservation, civic responsibility and consumerism. Lowe and Morrison (1984: 80) even go so far as to contend that a major attraction of environmental issues for the media is that they can be depicted in non-partisan terms, allowing journalists to subversively foster environmental protest at the same time as appearing to maintain a politically balanced stance.

Cottle (1993: 128) echoes Habermas in noting how the media debase the public sphere, refracting the environment through a journalistic prism that reduces politically charged stories such as global warming to the more immediate and mundane domestic and leisure concerns of ordinary consumers; for example, whether a beach holiday is likely this summer.

Constructing 'winning' environmental accounts in the media

As Stallings (1990: 88) has noted, some media accounts of environmental problems drop by the wayside while other 'winning accounts' persist and ultimately succeed in gaining acceptance. Indeed, the media contribute to this by fostering an image of either growth or decay for a particular problem (Downs 1972).

McComas and Shanahan (1999) have interrogated this notion of 'attention cycles'. The researchers content-analyzed stories in the *New York Times* and *Washington Post* from 1980 to 1995 about global warming. McComas and Shanahan found that narratives about this environmental issue passed through five stages: a pre-problem stage, a period of alarmed discovery, public realization of significant progress, gradual decline of intense public interest and a post-problem phase. Narratives about the implied danger and consequences of global warming were more prominent on the upswing of newspaper attention, whereas those dwelling on controversy among scientists received greater attention in the later stages (Dispensa & Brulle 2003: 93). Claims-makers thus need to learn how to keep environmental stories fresh and compelling.

In charting the ascent and tenure of environmental problems on the media agenda, it is possible to identify five key factors.

First, in order to gain prominence, a potential problem must be cast in terms that 'resonate' with existing and widely-held cultural concepts (Kunst & Witlox 1993: 4). This is why the frame alignment process is so crucial. Despite over three decades of exposure to environmental discourse, the actual awareness and salience of most environmental issues remains 'pitifully low' (Cantrill 1992: 37). In particular, most citizens, especially in North America, continue to place their faith in science and technology and to believe that economic growth is generally desirable. Thus, packaging an issue in the form of direct criticism of the Dominant Social Paradigm would not appear to be an effective communication strategy for environmental claims-makers. Instead, it makes more sense to situate environmental messages in frames that have wider recognition and support in the target population: health and safety, bureaucratic bungling, good citizenship and so on.

Second, a potential environmental problem must be articulated through the agendas of established *authority fora* (Hansen 1991: 451), notably politics and science. If it does not receive this legitimation, a problem will likely stagnate outside the media arena. This was the case in Britain where various 'green' issues (acid rain, ozone damage) lay relatively fallow until invigorated by a speech from Prime Minister Margaret Thatcher to the Royal Society in September 1988,[8] in which she adopted an environmental rhetoric for the first time. The Thatcher speech conferred a new degree of political legitimacy on the environment and the environmental movement and this subsequently diffused throughout many other arenas with the assistance of the mass media (Cracknell 1993).

Third, environmental problems that conform to the model of a publicly staged social drama are more likely to engage the attention of the media than those that do not. As Palmlund (1992: 199) suggests, the societal evaluation of risk takes the form of a dramatic contest coloured by emotions and containing both blaming games and games of celebration. In this contest there are readily identifiable heroes, villains, victims and even a chorus. Love Canal was the perfect media story by this yardstick with the timid housewife turned activist Lois Gibbs as the heroine, neighbourhood children with their increasing health problems as the primary victims and Hooker Chemical as the odious polluter.

Some environmental organizations, notably Greenpeace, have been very successful in staging morality plays in front of the global media with themselves as intrepid idealists and a changing cast of characters – whalers, seal hunters, French sailors, nuclear operators – as the villains. By contrast, problems that lack this fairy-tale quality, for example, the seepage of indoor radon gas into Canadian and American homes, are more difficult although not impossible to sell to the media.

Fourth, an environmental problem must be able to be related to the present rather than the distant future in order to capture media attention. Dianne Dumanoski, an environmental reporter for the *Boston Globe*, notes that some of the more immediate environmental problems such as oil spills interest editors more 'because they can understand that. . . . There's dirty stuff on the rocks; it's not computer models and these guys at MIT talking about something in the future' (Stocking & Leonard 1990: 41).

Global warming appeared to be a far away problem until the abnormally hot summer of 1988, when a series of tangible environmental disasters – droughts, floods, forest fires, polluted beaches – dominated the news. These contributed significantly to *Time* magazine's editorial decision to feature the endangered earth in its Planet of the Year issue of 2 January, 1989 (McManus 1989).

Finally, an environmental problem should have an action agenda attached to it either at the international (global conventions, treaties, programmes) or the local community (tree-planting, recycling) level. Environmental conditions that are less amenable to action are not as likely to appeal to reporters and editors unless, as was the case with the Ethiopian famine, a moral panic can be created around the consequences provoking a flurry of humanitarian relief efforts. Furthermore, rather than advocating some long-term action plan with results that may not be noticed for decades, environmental claims-makers should be able to offer the media some tangible results in the here and now: for example, shutting down an incinerator, cleaning up a polluted harbour, rescuing a beached whale. Unfortunately, as Solesbury (1976: 395) has noted, complex environmental problems with multiple dimensions are the most difficult to process because they can easily become bogged down in scientific disputes and interdepartmental rivalries. In such cases the media will tire of a problem, relegating it to a journalistic limbo where it is considered neither finally retired nor sufficiently topical to be of current public interest.

Mass-mediated environmental discourse

From a topic with no distinct identity of its own, the environment has progressed to the point where it is now an established part of everyday journalism. While there has been a broad upsurge in coverage, there is no single overarching environmental discourse. Instead, the media are the site of multiple outlooks and approaches, some of which are in direct conflict with the others (see Brulle 2000).

On one level, environmental communication is primarily an objectivist scientific discourse. As noted earlier in this chapter, journalists normally view themselves as impartial judges open to conversion only if the scientific proof is seen to be convincing. Scientific claims are reported at face value with relatively little attention given to their constructed nature, nor to their unknowns and uncertainties (Stocking & Holstein 1993: 202). Journalists have little patience with the thrusts and parries of scientific debate: either a danger exists or it doesn't.

At the same time, the media routinely lapse into a human interest discourse that 'carries the journalist out of the field of natural science and into the action oriented fields of social movements and politics' (Killingsworth & Palmer 1992: 135). Here, the burden of proof is less exacting. The essence of an environmental problem is more likely to be presented in a single dramatic image: a drum of toxic material, a discarded syringe on the beach, a head of foam on the surface of a trickling stream. Scientific skepticism is replaced by 'common sense'. The emphasis is less on the nature of the conditions that underlie the problem and more on the imputed consequences for people's lives. The narrative is more dramatic, even mythological.

Take for example a wire service story from the mid-1990s (Lawson 1994) on public hearings into a request by a joint Canadian–American venture to convert an unused oil pipeline running through rural Ontario (Canada) to a natural gas conveyance. Rather than examine the technical, economic and environmental feasibility of the project, the reporter chose to emphasize the participation as an intervenor of Jean Lewington, the widow of an area farmer who had spent 30 years successfully fighting a previous pipeline extension, thereby changing the way utility companies must deal with farmers and their land. This was accented by a photograph of Mrs. Lewington standing in front of her barn with a headline that read 'Farm widow re-fights old pipeline foe'.[9]

Third, the media, especially the business press, have increasingly adopted a discourse that presents the environment as an economic opportunity. The key message here is that environmental adversity can be turned into profit through human ingenuity and industry. Much of this type of coverage is product-oriented, touting a wide variety of 'green' products from the energy-saving house to nuts harvested by indigenous peoples in the rainforests of Brazil. The predominant message is that the entrepreneurial spirit need not be incompatible with ecological values; rather, the two are mutually reinforcing. This optimistic view of the environment has been amplified in the rapidly expanding body of stories on the promise and prospects for 'sustainable development'.

Fourth, the media situate the environment as the locus for rancorous conflict. While this environment as conflict package sometimes deals with the wider clash of cosmologies between environmentalists and their opponents, it is more likely to depict these disputes in the same manner as journalists routinely portray industrial relations disputes. Protesters are implicitly blamed for the disruption of normal commerce, the rationale for their actions is compressed into short sound bites and the background to the conflict is downplayed. The leaders of environmental protest actions are often presented as 'hippies' and violent 'ecoteurs' armed and ready for monkey wrenching[10] (Capuzza 1992: 12). Occasionally, however, journalists may choose to depict environmental protest in a more favourable light (see Box 5.2).

An environmental conflict story may shoot to the top of the news agenda if a well-known celebrity arrives on the scene. For example, the 1990s protest against clear-cutting the old-growth forest on Clayoquot Sound, Vancouver Island, which has been described as the 'war in the woods' that set the stage for today's battles over oil and gas pipelines ('Environmentalists mark milestone' 2013) was elevated in news value when Robert Kennedy Junior arrived to 'inspect the damage'. Rancorous environmental conflicts are supercharged with symbolic content with

---

Box 5.2 'Swampy'

One notable exception to the usual negative media coverage accorded environmental dissent is the treatment of youthful protesters against a road-building scheme on the A30 trunk road (highway) in Devon, England in 1997. In fact, a most unlikely hero known only as 'Swampy' arose from this protest. Swampy was the last of five protesters to emerge after camping for a week in a maze of tunnels underneath the road. Among other things, Swampy wrote a column in the *Sunday Mirror* for nine weeks; appeared on a popular TV news quiz comedy show and was the inspiration for a character in the long-running television soap opera *Coronation Street*. As Paterson (2000: 151) notes, Swampy 'became a byword for environmental direct action and youth disaffection from formal politics'. Paterson argues that the media de-activated the more radical elements in the campaign by normalizing Swampy and his fellow protesters ('Muppet Dave', 'Animal Magic'). For example, the *Daily Express* dressed Swampy in designer Armani suits for a photo shoot, and the *Daily Mail* profiled Animal Magic as a talented and articulate 16-year-old adolescent who even blushed when asked about her boyfriends. While this may have had the effect of making opposition to the road-building programme seem acceptably idealistic and legitimate (as against coverage of previous protest actions of this type elsewhere in the UK that were depicted as violent and extreme), at the same time it obscured 'the connections between road building and broader social and political questions and thus deep opposition of the road protesters to modern forms of organization and power' (Paterson 2000: 158).

---

both protesters and their opponents likely to use the framing and reasoning devices identified by Gamson and Modigliani (1989).

One consequence is the spillover of this media discourse into real life ideological battles between environmentalists and their opponents. Thus, Dunk (1994) observed that the forest workers in northwestern Ontario tend to regard environmentalists as outsiders from 'down south' or from 'big cities', in large part because they uncritically accept the dominant normative structure of the popular media's representation of environmental issues as a confrontation between middle-class, urban-based environmental radicals and local citizens fighting to keep their jobs.

Fifth, the media situate the environment within an apocalyptic narrative (see Chapter 1). Employing a series of medical metaphors, our planet is depicted as facing a debilitating, perhaps terminal, illness. Overpopulation, loss of biodiversity, rainforest destruction, ozone depletion and global warming are all linked causally to this impending ecological crisis. Despite the caution expressed in scientific media discourse, journalists give considerable news space to the popularized accounts of global threats formulated by Paul Ehrlich (overpopulation, biodiversity loss), Steven Schneider (global warming) and Norman Myers (tropical deforestation) and other prophets of doom. Thus *Time* magazine subtitled its 1989 special issue cover story on the greenhouse effect with the caption, 'Greenhouse gases could create a climatic calamity' (Killingsworth & Palmer 1992: 158).

Finally, the environment is scrutinized through the lens of institutional decision-making. Rather than attributing it a unique status, the environment is treated as just another policy area alongside health care, education and social services. The focus here is on regulatory agencies and processes, impending legislation, political personalities (Al Gore, Maurice Strong) and international fora (United Nations, European Community). Too often this leads to an ingrown policy debate between political and scientific elites (Wilkins & Patterson 1990: 21) in which the public is only an incidental bystander.

At any one time, various media packages as well as a plethora of individual news frames may compete for dominance. A single environmental event may have multiple shifting frames as it develops. For example, Daley and O'Neil (1991) trace the odyssey of the *Exxon Valdez* oil spill from a disaster narrative (the public as helpless victims, a catastrophe outside human control) through a crime narrative (the captain was culpable) to an environmental narrative (environmentalists contested the statements and practices of industry and government officials). At the same time, attempts to frame a story may fail. In the *Exxon Valdez* case a competing subsistence narrative (the oil spill posed a threat to native Alaskans' way of life) was all but ignored, appearing only in an indigenous publication, the *Tundra Times*. Journalists are thus faced with choosing from an assortment of narratives, languages and viewpoints at the same time as adhering to the formats and structures imposed by standard journalistic practice.

Conclusion

What should be evident from the discussion in this chapter is the considerable extent to which environmental news is socially constructed. In large measure this is a reflection of the rhythms and constraints inherent in the practice of journalism itself. In addition, it reflects the multiple competing claims that news workers must routinely sort out in the course of putting together a story. This central difficulty in reporting has been summed up by Stocking and Leonard in this way:

> The environmental story is one of the most complicated and pressing stories of our time. It involves abstract and probabilistic science, labyrinthine laws, grandstanding politicians, speculative economics and the complex interplay of individuals and societies. Most agree it concerns the very future of life as we

know it on the planet. Perhaps more than most stories it needs careful, longer-than-bite-sized reporting and analysis now.

(1990: 42)

Whether this depth of coverage is realistically possible is an open question that depends on several factors.

First, editors and producers, the news workers (and gatekeepers) who effectively set everyday lineups and assignments must see environmentalism as more than a transient phenomenon that loses its lustre once it ceases to register strongly in public opinion polls and government agendas. This is less likely to be the case in regions of the country where environmental conflict is endemic because of a natural-resource-based economy. Ironically, the one section of the media where environmental coverage has become institutionalized is in the financial pages, where 'green business' is seen as having increased relevance.

Second, environmental issues must be perceived as occupying a distinctive story niche rather than simply overlapping a multitude of existing subject areas – politics, business, agriculture, science and technology. Without a distinctive image, environmental coverage is destined to always remain event-driven and conflict-oriented. At the same time, environmental problems are by their very nature intricately tied in to economic and political structures and policies, making it difficult and sometimes even inadvisable to consider them separately; for example, this is the case with many Third World 'sustainable development' stories. It is thus difficult to balance the need for a distinct environmental specialty beat with the need for a depth of coverage that may reside in other areas of journalistic expertise.

Finally, some way must be found to combine 'muck-raking' or 'exposure journalism' with the longer-term goals of environmental education and policy reform. Investigative reports in the press or on television programmes such as *60 Minutes*, *Frontline* or *The Fifth Estate* may temporarily shock audiences but they do not necessarily result in either a deeper understanding of an issue or in effective regulatory action. Indeed, sometimes there can be a response quite different from that desired by activist claims-makers. Fletcher and Stahlbrand (1992: 195) cite what occurred in the early twentieth century when Upton Sinclair wrote a widely noticed exposure of the exploitation of immigrant workers in the large meat packing plants of Chicago in his book *The Jungle* (1905):

> His dramatic example of a man falling into a machine and being minced with the meat led not to a better protection for workers but rather to meat inspection laws, a reform the meat packers wanted to help them compete in European export markets.

In a similar fashion, a segment on *60 Minutes* concerning a community activist's fight against an incinerator, which, she charged, was emitting toxic pollutants, evidently resulted in a number of positive business enquiries from other American municipal governments to the waste management company that operated the facility. There must, then, be some blend of story elements that succeeds in raising an alarm in the public arena and then situating this concern within a clearly defined set of goals for environmental reform.

Notes

1　The phrase, 'Spaceship Earth' was evidently coined by the British economist Barbara Ward as the title of a book she published in 1966 on the links between economics and the environment (Pearce 1991: 11).

2　The only exception to this was the *New York Times* coverage that continued to separate various aspects of environmental issues.

3　Reporters' first choice here is usually a government spokesperson rather than a scientific expert. Sandman *et al.* (1987) suggest that one major reason for this is that reporters generally want two very specific types of environmental risk information: how much of the hazardous substance is in the air or water and how much of this substance does it take to cause problems.

4　Nearly 20 years earlier, an American researcher (William Witt) noted a similar diversity of environmental sources. Witt's results indicated that the primary news sources of environmental reporters were conservation clubs and organizations followed closely by business and industry sources (Witt 1974). It is worth noting that unlike Cottle, Witt did not extract his sources from media content alone, relying instead on a national questionnaire survey of environmental reporters working for US newspapers.

5　Einseidal and Coughlan (1993) found some revealing differences when they compared the environmental content in Canadian daily newspapers with full-time environmental writers with that in papers that utilized general reporters. On the whole, there were more environmental stories in the former; the environmental beat reporters were more likely to write longer, more analytical, self-initiated pieces and they were more likely to challenge conventional institutional wisdom.

6　According to a survey carried out by *Editor & Publisher* in the summer of 1970, there were 107 environmental reporters working in the American media, mainly on daily newspapers (Schoenfeld 1980: 456).

7　I witnessed this firsthand when doing observation in the newsroom of a national television network in Canada. One day, a senior producer was visibly upset when he received a letter from a viewer charging that the national news broadcast had been giving too much time to an anti-nuclear protest despite the newsworthiness of the issue (Hannigan 1985).

8　In his book *The Coming of the Greens*, published that same year, Jonathan Porritt, director of the environmental group Friends of the Earth, asked rhetorically, 'Is the Conservative Party totally immune to the sort of internal pressures within the other parties which allow one, however tenuously, to see the green tendencies emerging?' (Porritt & Winner 1988: 78). Thatcher's environmental turn may have had less to do with ideology than with politics. Determined to shut down the British coal mining industry and break the troublesome miners' union once and for all, Mrs. Thatcher announced in a speech to the United Nations the following year that the UK would establish a new centre for the prediction of climate change. At the same time, she strongly advocated the increased use of nuclear power generation (Yearley 1992: 20), a policy that was strongly opposed by the Green Party and its supporters and challenged the traditional dominance of coal as an energy source.

9   Not coincidentally, perhaps, the Lewingtons' daughter, Jennifer, became a beat reporter with the (Toronto) *Globe & Mail*.

10   'Monkey wrenching' or 'ecotage' refers to a wide range of actions by radical environmental activists to disrupt and halt damage to the environment including pouring abrasives into the crankcases of road-building vehicles, pulling up surveyors' stakes and 'spiking' trees by driving long metal spokes into them. The name comes from Edward Abbey's 1976 novel *The Monkey Wrench Gang* in which a group of 'ecoteurs' plots to blow up the Glen Canyon Dam (Franck & Brownstone 1992: 190; Manes 1990: 8–9).

Further reading

Anderson, A. (1993b) *Media, Culture and Environment*, New Brunswick, NJ: Rutgers University Press.

Boykoff, M.T. (2011) *Who Speaks for the Climate? Making Sense of Media Reporting on Climate Change*, Cambridge: Cambridge University Press.

Hannigan, J. (2012a) *Disasters Without Borders: The International Politics of Natural Disasters*, Cambridge: Polity Press.

Online resource

Environmental Media Association. Available HTTP: http://www.ema-online.org

6 Science, knowledge and environmental problems

It is rare indeed to find an environmental problem that does not have its origins in a body of scientific research. Acid rain, loss of biodiversity, global warming, ozone depletion, desertification and dioxin poisoning are all examples of problems that first began with a set of scientific observations. Ultimately, it is the scientific underpinnings of these environmental problems that lift them above most other social problems that are more dependent on morally-based claims (Yearley 1992: 117).

Furthermore, scientific researchers act as gatekeepers, screening potential claims for credibility. In 1988, when the British ecological organization Ark mounted a publicity campaign in which they alleged that melting ice-caps due to global warming would raise sea levels five metres in 60 years, thereby covering much of Britain with water, more sober scientific estimates of less than a metre rise quickly discredited the Ark initiative (Pearce 1991: 288–9).

Yet, paradoxically, science itself is frequently the target of environmental claims. One notable example of this is the contemporary debate over genetic engineering and its potentially harmful effects in the environment. In these cases, claims-makers explicitly reject the technical rationality of science in favour of an alternative cultural rationality that appeals to 'folk wisdom, peer groups and traditions' (Krimsky & Plough 1988: 107). Science is pilloried for interfering with the natural order rather than praised for lending its authority to a claim. The layperson's distrust of science and its users, Peuhkuri (2002: 159) suggests, may stem from past experiences where expert groups have appeared to be supporting powerful, trans-local interests at the expense of the local identity and way of life.

Science as a claims-making activity

The profile of science presented so far would seem to suggest that scientific findings reflect the physical reality of the natural world in a relatively straightforward manner. Science would therefore appear to be a search for truth in which the goal is to obtain a clear reflection of nature, as free as possible from any social and subjective influences that might distort the 'facts'. Yet, to the contrary, the assembly of scientific knowledge is highly dependent on a process of claims-making. In this regard, Aronson (1984) identified two types of knowledge claims made by scientists: cognitive claims and interpretive claims.

Cognitive claims aim to convert experimental observations, hypotheses and theories into publicly accredited factual knowledge. Blakeslee (1994) describes this conversion process as one in which scientists must adeptly stake novel claims while at the same time fitting them into an established research tradition. She gives as an example the process of cognitive claims-making in the physics journal, *Physical Review Letters*, in which contributors' letters announcing innovations have come to resemble journalistic accounts of scientific findings complete with an arsenal of rhetorical strategies.

Interpretive claims, on the other hand, are designed to establish the broader implications of the research findings for a non-specialist audience. Interpretive claims implicitly ask lay audiences to certify the social utility of the research, and the content of the claim supplies the reason they should do so. For example, in the case of global warming, the cognitive claim is that gases from cars, power plants and factories are creating a greenhouse effect that will boost the temperature significantly over the next 75 years or so. The interpretive claim here is that this heating trend is potentially dangerous because, among other things, it will cause havoc with the existing geography of the earth, flooding some low-lying areas such as the Netherlands and New Orleans and bringing drought to fertile agricultural regions such as the American Midwest.

Not only do scientists make knowledge claims, they also routinely construct 'ignorance claims' (Smithson 1989). This means that researchers highlight 'gaps' in available scientific knowledge in order to make a case for further research funding or, conversely, to retard further policy action on the grounds that not enough hard data exist to justify regulation or legislative activity (Stocking & Holstein 1993).

Aronson (1984) outlines three types of interpretive claims that scientists make: technical, cultural and social problem.

Technical interpretive claims-making occurs when researchers act as scientific advisers to industry and government. This often involves the evaluation of risks posed by controversial technologies (nuclear power, genetic engineering), suspected toxic pollutants (dioxin, mercury) and global hazards (ozone depletion, global warming). While in theory scientific advisers are restricted to a narrow technical assessment role, in reality they incorporate their own political agendas and knowledge claims into their own interpretations and recommendations.

Salter (1988) uses the term 'mandated science' to refer to the science that is used for the purposes of formulating public policy including studies commissioned by government officials and regulators to aid in their decision-making. Despite an official face of neutrality flowing from scientific expertise, members of expert panels regularly make moral and political claims and choices. These choices are fashioned as much by policy considerations as by scientific norms. For example, a scientific advisory committee dealing with pesticide safety may be equally aware that banning a chemical compound will negatively affect a US$500 million industry, while recommending its use could have serious health effects that will only become evident ten years later. This knowledge, Salter observes, affects the committee's recommendations, as does their technical data, thereby imbuing their activities with a strong interpretive flavour.

Cultural interpretive claims attempt to develop ideological support both for expenditures on scientific research and for the autonomy of science. The media through which the claims are presented are public speeches, articles in popular scientific magazines (*New Scientist*, *Scientific American*) and on the op-ed pages of influential newspapers (*New York Times*, *Washington Post*, *The Times* [London]), testimony before parliamentary enquiries and participation in government – industry committees and panels. In some cases, the receipt of an international scientific prize allows the researcher a unique platform from which to address broader social and political concerns. This is what occurred in Canada when John Polanyi won a Nobel Prize in chemistry and took advantage of the outpouring of public attention to address a raft of issues from government underfunding of universities to nuclear disarmament and peace. In other cases, the threat of a public

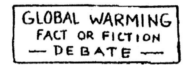

Figure 6.1 Global warming fact or fiction debate
Source: *Saturday Evening Post* 264(3), September/October, 1992.

review of scientific work can mobilize scientists towards making cultural interpretive claims. For example, Krimsky (1979) has demonstrated that the threat of external intervention and control into recombinant DNA molecule research in the 1970s turned American scientists into surprisingly effective lobbyists for scientific autonomy and the freedom of self-regulation.

Social problem interpretive claims assert the existence of a social problem for which a particular scientific specialty is uniquely equipped to solve. Aronson identified three conditions under which scientists are likely to make such claims.

The first is when a new discipline has no foothold in the academic world and therefore must appeal to external constituencies to obtain funding and political support for its work. To a degree, this has been the case for environmental science, which has been routinely criticized by many mainstream scientists for doing research that is defensive or of low quality (Rycroft 1991).

The second condition is when enterprising scientists, ever in search of new publicly derived research funds, attempt to show that their existing research work contributes to the solution of a recognized social problem or that it will successfully solve a previously unrecognized problem. This was characteristic of cancer research in the 1970s and AIDS research in the 1980s.

A third condition under which social problems claims-making is likely to occur is when scientists are confronted by social movements that seek to restrict or contain their research. In this situation, scientists are compelled to assemble and promote their own set of interpretive claims to either justify why a problem exists and their research should continue or why their research should not be construed as constituting a problem.

Aronson argues that there is a tendency for the first two forms of interpretive claims, technical and cultural, to eventually be transformed or subsumed by the social problem form because what is basically at stake is the social utility of science. Researchers realize that it is better strategy to proactively make a case for the social benefits of their work rather than wait and subsequently have to justify it in an atmosphere of skepticism and budget slashing.

Scientific uncertainty and the construction of environmental problems

What particularly opens the door to the creation and contestation of environmental problems is the inability of science to give absolute proof – unequivocal evidence of safety. Instead, scientists are reduced to offering estimates of probability that often vary widely from one to another. This lack of certainty allows claims-makers both within and outside science to assert that the situation is alarming, that the risk is too high and that society should do something about it.

Furthermore, mainstream science and green activists differ fundamentally as to when human intervention is necessary to protect the environment. This difference in perspective is nicely illustrated in a debate that took place in the early 1990s in the pages of the British science magazine, *New Scientist*.

Here, Brian Wynne and Sue Mayer argue that the decision whether to take official action on environmental risks should be governed by a *precautionary principle*. This states that if there is reason to suspect that a particular substance or practice is endangering the environment then action should be taken even if the evidence is not ironclad. The rationale behind this view is that it will be too late to respond effectively if we wait for a final scientific resolution years down the road. Where the environment is at risk, there is, they argue, 'no clear cut boundary between science and policy' (Wynne & Mayer 1993: 33).

The precautionary principle has been enormously influential, especially in Europe. British sociologist Adam Burgess (2003: 105), who views the concept as problematic, nonetheless acknowledges that it 'forms the basis for much domestic and international policy making' and, in its harder form, 'represents a frontal challenge to the experimental method that has been so central not only to science, but to modern society in general'. Theofanis Christoforou (2003: 205–6), a legal advisor to the European Commission, observes that in the EC, the precautionary principle has the status of nothing less than 'a mandatory treaty principle'. If properly applied, he says, it 'can be deployed to ensure that the societal values and democratic policy choices on health and environmental protection are fulfilled'.

Alex Milne, a consulting chemist who spent 34 years working in the paint industry, presents the contrary position. Milne rejects the precautionary principle, which he labels as one of the central doctrines of 'green science', as entirely the wrong approach. It is worse, he claims, than the legal principle in *Alice in Wonderland*, where the pattern was 'sentence first, verdict afterward'; here it is 'verdict first, trial afterward and no need for evidence' (1993: 37). The precautionary principle, he complains, has nothing to do with science: it is entirely an administrative and political matter.

A large measure of the disagreement here revolves around how science should be done. In traditional science, a reductionist principle predominates. This means that researchers break down a problem into the smallest number of constituent parts and look at each part separately, controlling as much variation as possible. If you want to look at the effect of a toxic chemical on the breeding pattern of fish, you isolate the fish in an experimental setting, vary the levels of the chemical and record the birth results. By contrast, a cardinal principle of green science is the necessity of looking at the world holistically. Since everything is connected to everything else, it does not make sense to disassemble an ecological web experimentally. For example, immunity is a complex system that is linked to a variety of factors from genetics to environmental pollution to socio-psychological stress. Causation may be indirect or multiple, making it all but invisible to the reductionist perspective of traditional 'good science' (Wynne & Mayer 1993: 34).

In policy terms, good science manifests itself in the form of an *assimilative approach*, which purports to define scientifically the capacity of an ecosystem to assimilate pollutants without harm and then to license industrial discharges within these 'proven' safe limits. What this ignores, environmentalists maintain, is the possibility of a chemical interaction among the polluting chemicals that creates a potential for end effects not anticipated by the assimilative model.

In evaluating the use, neglect and possible misuse of a precautionary approach to fourteen detailed cases of occupational, public and environmental hazards, Harremoës et al. (2002:187) conclude that a central lesson is that there will always be inevitable surprises or unpredicted effects, no matter how sophisticated knowledge is. For example, in the case of stratospheric ozone depletion, chemicals that were relatively inert under 'normal' conditions turned out to behave very differently under conditions that were not considered in the risk appraisal. A key element in a precautionary approach, then, is to acknowledge the possibility of surprise. Pragmatically, this requires greater care and deliberation in decision-making. The editors also recommend 'a broadening of appraisals to include more scientific disciplines, more types of information and knowledge, and more constituencies'.

As Salter (1988) has observed, quite different sets of criteria are applied depending on the context in which research evidence is evaluated. Conventional science possesses a deeply ingrained capacity to handle ambiguity; indeed, most journal articles routinely end with the caveat 'further research is needed'. By contrast, the burden of proof is stricter when scientists appear before regulatory hearings or in the courtroom. Here, legal concepts such as 'reasonable doubt' are prominent – anathema to scientists who are socialized to always couch their conclusions in conditional terms. In this regard, Yearley (1992: 142) points out that scientific expertise depends on elements of judgement and craft skill, informal aspects of science that can be highlighted in a legal or regulatory hearing to make scientific evidence appear like mere opinion. This tendency is even further exaggerated when environmental groups communicate using a moral discourse in a setting where the conventions of a scientific, legal or regulatory discourse predominate. The precautionary principle is a good example of an environmental principle that operates on a different plane of certainty than do societal control institutions.[1]

The crucial dilemma, then, is that social problem interpretive claims that rest on sound scientific evidence are generally more 'robust' than those claims only supported by opinion (Yearley 1992: 76), but there is a fundamental disagreement between environmentalists, scientists, regulators and legalists over what constitutes sound scientific evidence.

Blowers (1993) observes that scientific evidence is problematic as a basis for environmental policy-making in five ways. First, there is the problem of cause and effect that we have been discussing; this makes it difficult to establish responsibility for the externalities produced by polluting activities. Second, there is the problem of forecasting impacts; for example, the uncertainty about the incidence, distribution, timing and effects of global warming. Third, uncertainty over the consequences of present actions and the risks imposed on future generations may lead to a paralysis of policy or to a tendency to discount the future risks of present action. Sometimes, in fact, another future focused scenario – the crushing burden of a spiralling national debt – may discourage taking bold ameliorative or prophylactic steps in the here and now. Fourth, the frequent absence or sparseness of environmental data not only makes it more difficult to provide sound scientific judgements but it opens the door to manipulation by vested interests who claim that environmentalists have exaggerated the danger. Finally, the fragile interpretations of environmental science can easily run aground on the shoals of politics where conflicts between interests dominate. This is especially the case where one is dealing with broad speculative ideas such as the Gaia hypothesis rather than narrower, more empirically captured linkages.

Some environmental commentators have argued that science has fallen into a trap by relying on mathematical modelling, particularly large complex computer models rather than observation data to make major speculations about the future. Kellow (2007:177) calls this virtual environmental science 'dangerous for both the conduct of science and for the use of science as the basis of public policy-making'. Examples include the *Limits to Growth* study (see Chapter 1), estimates of species extinction in conservation biology (see Chapter 8) and climate modelling, which relies on historical reconstructions.

Identifying environmental problems as scientific issues

Seldom does an environmental problem pop up overnight with no past legacy of scientific observation and debate. Rather than grow along a linear path, the process by which environmental problems are identified and evolve as scientific issues is characterized by the creation of a pool of knowledge that expands serendipitously in unexpected directions (Kowalok 1993). Individual pieces of data in this pool may be generated through projects that employ the reductionist methods of traditional science, but in the end it is a flash of holistic insight that leads to final understanding.

Despite appearances to the contrary, the basic outline of many environmental problems has been around for a long time. For example, the theory that greenhouse warming is caused by human-generated emissions of carbon dioxide has been known for more than a century, but the greenhouse effect was not considered a priority problem until the 1980s (Cline 1992: 13–14). Similarly the term 'acid rain' together with many of its fundamental principles was first introduced by chemist Robert Angus Smith in 1872 but did not emerge as a full-blown scientific problem until the 1970s.

What then propels an environmental problem of long standing into a current scientific claim of critical proportions?

First, the real or perceived magnitude of the condition may suddenly rise to 'crisis' proportions. For example, species extinctions have been occurring at a modest rate since 1600 as human settlements have spread across the globe.[2] Recently, however, environment movement organizations have alleged that we are seriously tipping the balance between the appearance of new species and the extinction of existing ones (Tolba & El-Kholy 1992). At the same time, the loss of old growth forests and plant and animal species captures the attention and concern of conservation biologists and other scientific claims-makers precisely because these natural resources are said to be down to their last 20, ten or one per cent, making preservation appear more crucial.

Second, new methodologies, research instruments or data banks may allow scientists to come to conclusions that were impossible earlier on. For example, data provided by the European Air Chemistry Network starting in the 1950s allowed Swedish researcher Svante Oden to advance his pioneering theories about acid rain, while James Lovelock's comparisons of the concentrations of fluorocarbons in the lower atmosphere with annual amounts of industrial production opened the door to chemists Mario Molina and Sherwood Rowland to document the key link between CFC products and ozone destruction (Kowalok 1993).

Third, the holistic character of global ecosystems means that rising scientific and public interest in one environmental problem readily generates interest in another interrelated problem. Thus, scientific concern over tropical deforestation has spread well beyond the boundaries of silviculture (a branch of forestry dealing with the development and care of forests) due in large part to the key role that the loss of tropical forests plays in what are presently the two highest profile global environmental problems: global warming and the loss of biological diversity. Mazur and Lee (1993) illustrate this in schematic fashion, demonstrating how the rise of public concern over the problem of the global environment is actually a weaving together of several strands of concern over specific problems, each of which has arisen at a different point in time. This synergy is not, of course, always readily apparent and scientific entrepreneurs may need to explicitly establish the relevance of one issue for another.

Fourth, the establishment of official research programmes, centres and networks may create a hothouse in which research into an environmental problem may be successfully nurtured, even if this is not the original intention. For example, the decision in December 1979 by the Council of the European Community to establish a multiannual research programme in the field of climatology was taken in part because of concern about what was essentially a regional problem – the 1976 drought that affected some African and European areas. Once in place, this programme became both the focus of foundation-building research on the physico-chemical processes related to increasing concentrations of greenhouse gases in the atmosphere and a source from which scientific findings and terms such 'greenhouse effect' and 'climate change' circulated outwards into EC policy-making circles (Liberatore 1992).

In all of this, the identification and characterization of threats is highly dependent upon a network of international scientific conferences and collaboration (Kowalok 1993: 36–7). Not only does this permit researchers to learn new methodological techniques or to find the missing pieces in their own puzzles; it helps build their confidence that they are not alone, an especially important shot of morale boosting when a theory seems radically new and controversial. This was very much the case with the research on the acid rain problem where Canadian and American researchers did not fully appreciate the global relevance of their own findings until they came face to face with similar findings from Scandinavia as presented by Svante Oden on his 1971 lecture tour of North America (Cowling 1982).

Coming out: communicating new environmental problems to the world

The transition from cognitive to interpretive scientific claim is comparable to a 'coming out' in which the ingénue makes a public representation of identity. At some point, the circulation of information around an essentially closed scientific loop is interrupted and the urgency and salience of a problem is shared with the outside world.

One common way of doing this is to convene a public forum at which a mixture of scientists, environmentalists and administrators jointly address the various dimensions of the problem in the full glare of the media spotlight. Alternatively, a claim may be articulated at a congressional or parliamentary hearing where media coverage is usually assured. For example, the 1981 US Congressional testimony by Peter Raven and Edward Wilson was important in establishing the economic utility of preserving endangered species of insects such as the butterfly or the honeybee, particularly for the development of new crops, pharmaceuticals and renewable energy sources (Kellert 1986). Similarly, the ozone depletion issue in Britain was not launched until parliamentary hearings were held in early 1988; strong representations were made in both Houses of Parliament to the effect that the United Kingdom must become a world leader in the drive to protect the ozone layer (Benedick 1991). A third channel for the dissemination of newly constructed scientific environmental problems is a scholarly conference at which reporters from major newspapers are present looking for 'blockbuster' theories. This is what occurred in September 1974, when the *New York Times* picked up on a delivered paper dealing with the threat of CFCs to the ozone layer; the *Times* article 'signaled the beginning of public concern over CFCs and their use in aerosol cans and refrigerators' (Kowalok 1993: 19).

In other cases, however, this process is short-circuited when scientific entrepreneurs go directly to the media. Svante Oden, the Swedish soil scientist who first proclaimed the theory of acid rain, published an account in the Stockholm newspaper *Dagens Nyheter* a year before he published in a scientific journal and five years before the issue arose at the 1972 UN Conference on the Human Environment. Similarly, in Germany, biochemist Bernhard Ulrich's hypothesis that huge tracts of German forests would be dead within five years due to damage from acid rain was presented as established fact in an article in *Der Spiegel*, a mass circulation periodical, provoking widespread national alarm.

How effective one channel is compared to another depends on a number of factors. If there is no consensus among scientists themselves and strident opposition from industry, a more individual approach may work best. Despite periodic attempts to raise the issue, the problem of pesticide poisoning in the United States was being suppressed[3] until Rachel Carson (1962) published her indictment in *Silent Spring*. Subsequently, a number of scientists came forward in her defense. The problem was legitimated when, in May 1963, a special panel of the President's Scientific Advisory Committee released a report that was critical of the pesticide industry. On the other hand, jumping the gun before scientific consensus has been established may succeed in capturing media and public attention but at the risk of bringing peer censure by fellow scientists. This occurred in 1988 when James Hansen, director of the NASA Institute for Space Studies, testified before a Senate that heat waves such as those being experienced at the time were directly attributable to the greenhouse effect. This norm within science against premature revelation has no doubt been strengthened as a result of the 'hoax' over cold fusion in which the researchers announced their findings at a press conference in Utah prior to subjecting them to peer review.

Science and environmental policy-making

In order for a scientific issue to become policy it must be translated into something that is 'treatable'. As a result, at the policy formulation stage the contribution of natural scientists usually diminishes while the role of socioeconomic and technical experts grows. For example, Liberatore (1992) found that while natural science findings still played an important role in the international debate on global warming in the early 1990s, it was the input of economists, policy analysts and energy technology experts that was crucial in shaping the nature of the European Community response.

Political scientists have captured this relationship between science and policymaking by using two concepts: epistemic communities and policy windows.

Epistemic communities

Haas has described the contribution of *epistemic communities* as critical in achieving international cooperative agreements on environmental issues. Epistemic communities are 'transnationally organized networks of knowledge based communities'; that is, internationally linked groups of specialists who offer technical advice to political decision-makers.

What gives them a key role in a process usually closed to non-politicians is the uncertain nature of environmental problems. Political leaders may be highly skilled in negotiating trade pacts or treaties but they feel at a distinct disadvantage in dealing with planet-threatening conditions relating to atmospheric shifts or chemical overloads. Under such circumstances, information is at a premium as a strategic resource, and politicians, in order to reduce such uncertainty, 'may be expected to look for individuals who are able to provide authoritative advice on whom to pin the blame for a policy failure or as a stop-gap measure to appease public clamour for action' (Haas 1992: 42).

Epistemic communities, Haas contends, are not only bound together by a technical expertise; they also share a number of causal and principled beliefs. In the case of environmental issues, these communities of knowledge were initially comprised of ecologists who share a belief in the need for a holistic analysis – a view that carries over to the policy advice that they give. For example, this was characteristic of an epistemic community of ecologists and marine scientists who spearheaded intergovernmental efforts in the 1980s to control pollution in the Mediterranean Sea (Haas 1990).

An epistemic community has the capacity to be influential both in defining the dimensions of a problem and in identifying likely solutions. Thus, Haas demonstrates how a transnational epistemic community of atmospheric scientists succeeded in influencing the negotiations that led to the signing of Montreal Protocol on the protection of the ozone layer in 1987 by 'bounding discussions on the broad array of substances to be covered and the rapidity of regulations' (Haas 1992: 49). Once the epistemic community has laid out the basic parameters of the settlement, it is up to the political leaders to decide what compromises have to be made in order to obtain agreement.

One especially influential conduit through which an epistemic community can shape the policy process is the international scientific assessment panel. Citing the examples of the Millennium Ecosystem Assessment, the Global Biodiversity Assessment and the Intergovernmental Panel on Climate Change (IPCC), Mooney (2003: 49) identifies five features that give these fora high credibility: (1) these panels carefully evaluate peer-reviewed literature; (2) they usually provide some measure of the certainty of the conclusions they draw; (3) the participants are balanced in expertise, region and gender; (4) the results of the assessment undergo rigorous review at many levels and (5) the final document puts the technical findings into terms that are relevant to policy.

These assessment panels, however, are not necessarily fully representative of all researchers or of the full spectrum of scientific claims pertaining to a particular controversy. For example, the IPCC, whose considerable importance is that it provides the scientific consensus and legitimacy that underpins the Kyoto Protocol, is said by some to disproportionately favour the views of the climate-modelling community found in a handful of large research laboratories often associated with meteorological offices (Boehmer-Christiansen 2003: 81).

Not only do these modellers differ from other researchers working on the global climate change issue, but also they vary internally as well. In his ethnographic study of the *epistemic lifestyles* – the strategies and assumptions they use to build and validate models of the climate – Simon Shackley (2001) demonstrates that there is considerable variation in the modelling styles and therefore in the kinds of knowledge claims associated with differing national and laboratory cultures (Miller & Edwards 2001: 20). These results show that scientific certainty cannot be divorced from epistemic lifestyle but, instead, must be 'negotiated'. The existence of epistemic lifestyles, Shackley concludes, suggests one important source of diversity in the practice of climate science and indicates that agreement on the role of human-induced global warming may be somewhat less uniform than is often assumed.

Not all political analysts agree with the elevation of Haas' scientific coalitions to a central point in the environmental decision-making process. Haas' model is said to break down in the degree of autonomous power accorded to the epistemic community. That is, scientific coalitions can use their resources to highlight a problem but they must enlist political leaders from their individual nations to have a real impact on treaty negotiations. These leaders may find it advantageous to engage in international problem solving but ultimately they are guided by domestic political considerations (Susskind 1994: 74–5).

Individual governments depend on the technical expertise built up by environmental movement organizations such as Friends of the Earth, Greenpeace and Pollution Probe. In recent years, these groups have devoted considerable resources to building up their own in-house research capabilities, hiring scores of bright, young, idealistic Ph.D.s fresh out of graduate school. In addition, conservation and environmental organizations typically have scientific advisory committees and call upon the voluntary support of university scientists and civil servants who are scientists (Yearley 1992: 126). As a result, there is a synergy between organizations and official policy-makers who find the knowledge and information produced by Greenpeace and others to be of considerable value in staking out their position in public arena debates over environmental issues (Eyerman & Jamison 1989; Lowe & Goyder 1983).

While epistemic communities may be international in scope, the centre of gravity for scientific claims-making on specific issues tends to reside in a specific nation. For example, it was US scientific leadership that propelled the ozone depletion problem into global prominence, while Swedish (and Norwegian) research on acid rain was vital in elevating that issue to problem status. In the former case, a critical infrastructure clearly existed as the result of the space programme and the pre-eminence it gave to the United States in researching the stratospheric

Box 6.1 The 'hockey stick'

In making the case that human activity in the industrial era is primarily responsible for global warming, one very powerful promotional tool has been a graphic nicknamed the 'hockey stick'. Like the 'hole' in the ozone layer, this commands public attention by presenting a visual image that is easy to identify and recall. The graph is a reconstruction of temperatures over the past millennium, assembled from data from tree rings, corals and other markers. For most of this period, there are evidently only relatively small fluctuations in temperature (the stick shaft); then, at the beginning of the twentieth century, there was a sharp upward movement (the blade of the stick).

First published in a 2001 report by the United Nations Intergovernmental Panel on Climate Change (IPCC), the hockey stick graph has been replicated in presentations and brochures used by hundreds of environmentalists, scientists and policy makers (Regalado 2005: A1). The Canadian government, for example, promoted the hockey stick on its website, sent it to schools across the country and cited it in pamphlets mailed out to all Canadians (McIntyre 2005: FP 19).

However, the hockey stick graph has come under sustained attack. This critique emerged initially from an unlikely source: Stephen McIntyre, a semi-retired Canadian financial consultant to small minerals-exploration companies, who claims he found some basic mathematical errors in the model. Together with Ross McKitrick, a Canadian economics professor and fellow 'climate skeptic', McIntyre began to publish his views, first in the British social science journal Energy & Environment, whose editor Sonia Boehmer-Christiansen is known for her maverick views on the global warming issue, and then in Geophysical Research Letters, a peer-reviewed scientific journal. Climatologist Michael Mann, lead author of the original article in Nature where the hockey stick graph was introduced, denounced the McIntyre-McKitrick critique of his work as 'frivolous' and politically motivated (Regalado 2005: A13). Nonetheless, Dr. Mann and his co-authors were compelled to publish a partial correction in Nature, although they still maintain that the data are sound.

While the majority of the international scientific community continues to endorse the IPCC Report, there are pockets of dissent. Some question whether the hockey stick graph significantly underestimates temperature fluctuations prior to the twentieth century, most notably in the years around AD1000, and again between 1400 and 1600. Others maintain that Dr. Mann and his colleagues erred in relying heavily on US bristlecone pine records, misinterpreting a hockey stick configuration as a temperature signal, rather than as evidence of a different biophysical trend. In reply, climate specialists who support Mann's conclusions point to a host of other indicators that the planet is warming up, from receding glaciers in Alaska to Mount Kilimanjaro, stripped of its snowcap for the first time in 11,000 years (Lovell 2005).

sciences. This was particularly located in two government agencies – NASA (National Aeronautics and Space Administration) and NOAA (National Oceanic and Atmospheric Administration) – as well as in the graduate faculties of major American universities (California, Harvard, Michigan). When researchers at these institutions voiced concern about events in the stratosphere, the site of the ozone problem, the media and the general public as well as political leaders tended to pay attention (Benedick 1991). In the case of acid rain, the forests and lakes were seen as a vital component of the Swedish economy and therefore were accorded high research priority. When the transnational origins of acid precipitation became obvious in the research data reported by Oden and others, the Swedish government did not hesitate to aggressively present these findings at the 1972 Stockholm Conference.

Policy windows

Another political science model that can be used to link science and policy-making is Kingdon's 'garbage can' model. Adapted from a model of organizational choice developed by James March and his colleagues, this proposes the operation of three major process streams in government agenda setting: (1) problem recognition, (2) the formation and refining of policy proposals and (3) politics. These three streams usually develop and operate largely independently of one another. However, at critical times the three streams may come together or 'couple'. Kingdon describes this as the opening of a *policy window* and attributes the main responsibility for this action to 'policy entrepreneurs' within the political system. Individual entrepreneurs do not open the window but they take advantage of the opportunity once it has occurred. At key junctures, then, solutions become joined to problems and both are joined to favourable political forces.

Hart and Victor (1993) have employed Kingdon's model to explore the role of scientific elites in influencing American policy on climate change for the years 1957–74. In their interpretation, science, policy and politics evolve in separate unconnected streams creating both solutions in search of problems and political problems in search of solutions. Scientific elites, assuming the entrepreneurial role, identify policy windows and seize advantage of them.

This is what occurred in the United States in the 1970s. For the better part of 20 years, two interesting scientific discourses relating to the climate had been meandering along, attracting some support but unable to really get moving in terms of either funding or public recognition. These were the 'carbon cycle discourse', which addressed the question of whether and why atmospheric concentrations of carbon dioxide (CO_2) were increasing, and the 'atmospheric modelling discourse', which asked what would happen to the climate if higher concentrations of CO_2 were reached. The former discourse was coordinated by an oceanographer, Roger Revelle, whereas John von Neumann, the father of scientific computing, promoted the latter.

In the early 1970s, the rise of the American environmental movement created a policy window that these elite scientists successfully exploited in order to mobilize financial and political support and raise public awareness. Hart and Victor (1993: 661) describe this as a synergistic relationship in which scientific findings such as those relating to the greenhouse effect 'catalyzed the rebirth of environmentalism' while at the same time environmentalism 'acted as a midwife for new scientific agendas – legitimating them and providing constituencies for their results'. Especially influential in linking the two research streams was Carroll Wilson, an MIT management professor, who was the guiding spirit behind the publication in 1970 of a report, entitled *Study of Critical Environmental Problems*, which was explicitly interdisciplinary and environmentalist in tone.

Hart and Victor (1993: 668) emphasize that very little new scientific information about the prospects of global warming was produced between the late 1960s and the early 1970s. Rather, what was different was that the two lines of research were brought together in a new, redefined, scientific agenda that was then successfully sold to political decision-makers and to the news media as a global environmental 'pollution' problem. As we discussed in Chapter 3, this presentation will be enhanced if a simple, visual metaphor is utilized. The 'hole' in the ozone layer is one example of this. Another, more controversial, one is the 'hockey stick' graph (Box 6.1) that has been used to make the case that temperatures have been spiking upwards since the beginning of the twentieth century and this is evidence that human activity in the industrial era is causing dangerous global warming.

Scientific roles in environmental problem-solving

As Skodvin (2000a: 139) observes, scientific findings rarely speak for themselves; rather, the skill and power of individual knowledge brokers in bringing scientific findings to the attention of policy-makers is crucial. In the case of the Montreal Protocol on the protection of the ozone layer, brokers played a critical role in reducing scientific uncertainty by shifting the debate from a focus on ozone depletion to that of atmospheric concentrations (Litfin 1994). Knowledge brokers frequently come from within the scientific community itself.

Susskind (1994) has proposed five primary 'roles' that are played by scientific advisers in the environmental policy-making process: trend spotters, theory builders, theory testers, science communicators and applied policy analysts. These roles frequently overlap but each has its own tasks and agendas.

Trend spotters are scientists who are the first to detect changes in ecological patterns and to correctly understand their significance. Occasionally, the trend spotter may be a lone scientist who observes some important pattern in the micro-ecology of the pond or marsh and is able to extrapolate this on to the larger environmental canvas. More common, however, are trend spotters who are part of a scientific team that is engaged in gathering and analyzing longitudinal data such as that assembled from the LANDSAT satellite or from the European Air Chemistry Network.

Theory builders try to explain the causes for the changes that the trend spotters identify. They are inclined to engage in model building, both to fit explanations to past circumstances and to predict future effects.

Theory testers critically scrutinize the models suggested by theory builders. Using pilot tests or controlled experiments, they attempt to ascertain whether the hypotheses and propositions generated by the model can be empirically proven.

Science communicators attempt to translate difficult-to-decipher data into terms that the public at large can understand. They are key players in the 'coming out' process that was discussed in an earlier section of this chapter. Some communicators such as Edward Wilson are eminent scientists who feel a strong moral responsibility to bring the fruits of their research to the public. Others, for example the Canadian geneticist and broadcaster David Suzuki, are researchers who have made a conscious decision to spend their life popularizing science and carrying the ecological message to a wider audience.

Applied policy analysts act as consultants to political decision-makers, converting scientific findings into policy recommendations. They play a prominent role in the formulation of environmental treaties because they take what is often abstract scientific information and recast it in terms that are amenable to legislation or to international agreements.

Each of the five types of scientists may contribute throughout the environmental problem-solving process but there is a considerable degree of specialization; that is, trend spotters and theory testers are usually more prominent during the fact-finding stages while science communicators and policy analysts play key roles during the negotiation/bargaining period (Susskind 1994: 77). In terms of the three key tasks in constructing environmental problems discussed in Chapter 3, trend spotters and theory testers can be said to characterize the 'assembling' process, communicators in 'presenting' an issue and applied policy analysts in 'contesting' an environmental claim.

Regulatory science and the environment

One important arena in which environmental science interacts with politics is the regulatory process. The 'regulatory science' that is found here differs from conventional research science in a number of ways (Jasanoff 1990). First, it is done at the margins of existing knowledge where fixed guidelines for evaluation may often be unavailable. Second, it usually involves a greater degree of knowledge synthesis than research science does, which puts a greater emphasis on the originality of findings. Third, science-based regulation requires a hefty dose of prediction, especially with regards to risk creation.

Jasanoff (1990: 230) argues that a negotiated and constructed model of scientific knowledge 'closely captures the realities of regulatory science'. Rather than encouraging an adversarial process, regulatory agencies seek scientific input into their decisions as a means of legitimation. This often takes the form of an ongoing scientific advisory committee. Jasanoff reviews a number of cases in which such advisory boards played a key role in decision-making at the Environmental Protection Agency (EPA) in the United States. In the case of air pollution, the relationship between the EPA and the Clean Air Scientific Advisory Committee (CASAC) was initially rocky but after extensive negotiation was transformed into a fundamentally sympathetic orientation. Similarly, despite problems during the Reagan era, the EPA's agency-wide Scientific Advisory Board (SAB) was able to maintain a respected and autonomous position, in large part because it focused on issues pertaining to scientific assessment while leaving rule-making activities to the agency proper.

In this negotiated model of science, Jasanoff contends, there can be no 'perfect, objectively verifiable truth', only a 'serviceable truth' that balances scientific acceptability with the public interest. In this context, scientific reality is clearly socially constructed so as to conform to a societal mean. However, in circumstances where sharply conflicting constructions of science land at the feet of a scientific advisory committee, reconciliation can often be most difficult. This is what has occurred in various regulatory controversies involving agricultural pesticides where scientific evidence has been especially difficult to establish while public concern has been high. In these situations, the debate over the precautionary principle, which we surveyed earlier in this chapter, rears its head, with scientific advisers opting for the traditional reductionist position while agency staff are more sensitive to the public pressure to act sooner rather than later. Where this occurs, the risk debate can easily shift to the arenas of the media and politics where it will continue under a different set of ground rules from those encountered in the regulatory setting (Jasanoff 1990: 151).

Conclusion

In an ideal world, scientists operate at a safe distance from the slippery world of political wheeling and dealing depicted so effectively in the acclaimed television drama series *House of Cards*. The assigned role of the researcher is to report findings that can be counted on to be factual and reliable, leaving it up to elected politicians and their civil servants to embed these in policies and legislation. The power and authority of science derives not only from its monopoly over knowledge production, but also from its reputation as an honest reporter, telling it like it is no matter what political outcome might ensue.

This unblemished view of science and politics has never prospered particularly well in the real world. A major reason for this is that individual scientists become passionately committed to a course of action and feel compelled to promote and defend it in the political arena.

One of the first writers to bring this into the open was the British scientist and novelist C.P. Snow. In a series of lectures on government and citizenship delivered at Harvard University, Snow (2013 [1961]) told the story of a bitter rivalry between two leading British scientists during World War Two that pivoted on the question of how to best defeat Hitler's Germany. Sir Henry Tizard, a chemist from Imperial College, advocated investing in the new science of radar in order to win the air war. Frederick Lindemann (Lord Cherwell), a physicist from the University of Oxford, recommended the massive bombing of working-class homes to break the German spirit. In the lead-up to and first year of the War, Tizard's policy recommendations found favour with the ruling Labour government. When the Conservatives assumed power, Lindemann became personal adviser to Prime Minister Winston Churchill and his strategy won out. Snow makes the point that vital decisions about the air war were made in secret and with little regard to truth or the prevailing scientific consensus.

In similar fashion, environmental science has in recent years become highly politicized, especially when it comes to the issue of global climate change. Pielke (2004) observes that scientists are increasingly equating particular scientific findings with certain political and ideological perspectives. This presents a potential threat to the institutions of science and democracy. Pielke especially frowns on a *linear perspective* wherein scientists adhere to a 'get-the-facts-then-act' model. This linear model 'may simply mask normative disputes in the language of science, to the possible detriment of both science and policy' (2004: 409). He much prefers an approach whereby scientists offer up a range of alternatives to policy makers and leave it up to them to decide. Science, he says, 'never compels just one political outcome' (2004: 406).

Notes

1   An exception to this is Germany, where the precautionary principle has been enshrined historically.

2   The World Conservation Union estimates that there have been in excess of 800 plant and animal extinctions since 1500 when accurate historical and scientific records began. Putting this in perspective, Kellow (2007: 18) notes that this amounts to a documented rate of 1.6 extinctions per annum.

3   Scientific concern over pesticide poisoning began more than two decades prior to the publication of *Silent Spring*. As far back as 1945, Rachel Carson herself evidently attempted unsuccessfully to interest *Reader's Digest* in commissioning an article from her on the research being conducted by colleagues at the Paxutent Research Center indicating that the pesticide DDT had adverse effects on the reproduction and survival of birds after repeated applications (Lear 1993: 33). In the early 1950s, an emerging consensus in the US public health field that the use of chemicals in food production needed to be more strictly regulated led to 46 days of Congressional hearings. However, the issue was seen as narrow and technical and received little media attention. Unlike the eventual environmental campaign sparked by Carson's book, evidence that pesticides might cause harm somewhere down the road was not as compelling to the media and the mass public as dramatic images of dead birds (Bosso 1987: 80).

In the years 1957 to 1959 there were a series of pesticide-related accidents, notably massive fish mortality throughout New York State due to a gypsy moth spray campaign and the 'Great Cranberry Scare', in which cranberry sales fell by two-thirds after some of the fruit was found to be contaminated by residues of the herbicide aminotriazole. Yet these controversies were seen as being isolated and were not sufficient to change the *status quo*.

Further reading

Haas, P. (1990) *Saving the Mediterranean*, New York: Columbia University Press.

Jasanoff, S. (1990) *The Fifth Branch: Science Advisors as Policymakers*, Cambridge, MA: Harvard University Press.

Pielke Jr., R. (2004b) *The Honest Broker: Making Sense of Science in Policy and Politics*, Cambridge: Cambridge University Press.

Online resource

Hoffman, A. (2012) 'Climate change as culture war'. Talk to the Beadie School of Business, Simon Fraser University, October 23 (YouTube video). Available HTTP: http://www.youtube.com/watch?v=0kf1l3zyTmM

7   Risk construction

To an ageing population concerned about preventing heart disease, salmon has proven to be a tasty remedy, especially in summer when it can be grilled to perfection on the barbeque. Thanks to the worldwide growth of aquaculture, consumers can obtain farmed salmon year-round at relatively low prices. Eating 'oily' fish such as salmon twice a week, the American Heart Association tells us, confers the health benefits of Omega 3 fatty acids that can reduce the risk of sudden cardiac death following a heart attack.

Suddenly, in January, 2004, all bets seemed to be off when the respected journal *Science* published an article warning that farmed salmon contains alarmingly high levels of cancer-causing toxins, ten times more than in wild salmon. Risk analysis indicated, the authors warned, that 'consumption of farmed Atlantic salmon may pose health risks that detract from the beneficial effects of fish consumption' (Hites *et al*. 2004: 226).

As it happens, the *Science* piece was not the first research to come up with results of this type. Three years before, BBC News broadcast a programme, 'Warnings from the Wild, the Price of Salmon', that reported the results of a pilot project conducted under the auspices of the David Suzuki Foundation, that found farmed salmon had a higher level of PCBs and two others toxics than wild salmon. Then, in July 2003, the Environmental Working Group (EWG) in Washington released a report entitled 'PCBs in Farmed Salmon: Factory Methods, Unnatural Results'. The EWG bought salmon from local grocery stores in the United States. When the samples were analyzed in the lab, it was found that seven of ten fish were seriously contaminated with PCBs, raising concerns about cancer and fetal brain development. Based on their data analysis, the EWG concluded that 800,000 American adults ingest enough PCBs from farmed salmon to exceed the allowable lifetime cancer risk 100 times over.

Other health agencies and researchers hopped on the defensive. The US Food and Drug Adminstration (FDA) advised that the levels of pollutants found in salmon are too low for serious concern and urged Americans not to let the new research frighten them into a diet change. Eric Rimm, a specialist on nutrition and chronic disease at the Harvard School of Public Health, told the Associated Press that the *Science* article 'will likely over-alarm people in this country' (CNN 2004). It was pointed out that the study tested salmon raw with the skin on – removing it and grilling the fish removed a significant amount of PCBs, dioxins and other pollutants. One university toxicologist, an industry consultant, went so far as to venture that 'in my view, the study says we should be eating more farmed salmon' (Stokstad 2004: 154).

Despite this counter-offensive, salmon as a healthy meal choice had lost its lustre. Some consumers began to avoid salmon altogether. Others insisted on wild salmon, a questionable strategy since many stores and restaurants routinely sell farm-raised salmon as 'wild' (Burros 2005). At a restaurant dinner with sociology colleagues shortly after publication of the *Science* article, none of those present would even consider ordering the salmon, citing recent research that stated that this was 'risky'.

To a large extent, this episode is characteristic of how individuals in contemporary society engage in the processes of risk perception and assessment. Typically, we hear a brief item on the radio or see it in a newspaper or on the Internet, it comes from a seemingly reputable scientific source and it taps into an existing well of concern about our health or the safety of our family. This is true not only for food and lifestyle choices but also for risks related to technology and the natural environment.

Few of us have the time to systematically evaluate these risks, so we make snap decisions based on the inclusion in media reports of hot button phrases such as 'cancer causing'. Consider, for example, a 2012 Associated Press item with the tag line 'Samsonite pulls luggage from Hong Kong stores'. Evidently, the local consumer council in Hong Kong reported finding high levels of chemical compounds linked to cancer in the handles of the 'Tokyo Chic' line of Samsonite luggage. While insisting that the bags were completely safe, the company nevertheless recalled the product. In this case, there was no firm evidence to suggest that travellers risked falling victim to a cancer traceable to suitcase handles, only that the handles were manufactured with compounds classified in the laboratory as being carcinogenic.

Until recently, the published literature on risk almost uniformly reflected the belief that risks be 'objectively' determined, that this determination was exclusively the province of engineers, scientists and other experts and that any failure on the part of ordinary citizens to fully accept this was considered irrational. Risk assessment was thus conceived of as a technical activity where results were to be formulated in terms of 'probabilities'. There was even an emerging category of specialists – what Dietz and Rycroft (1987) have termed the 'risk professionals' – who make it their business to work out new methods of risk analysis.

Risk and culture

The first notable challenge to this position came from an eminent British social anthropologist, Mary Douglas, and an American political scientist, Aaron Wildavsky (both now deceased), who published a provocative book in 1982 entitled *Risk and Culture: An Essay on the Selection of Technological and Environmental Dangers*. *Risk and Culture* asks two simple but fundamental questions: why do people emphasize certain risks while ignoring others, and, more specifically, why have so many people in our society singled out pollution as a source of concern? The answers, Douglas and Wildavsky insist, are embedded in culture.

In their view, social relations are organized into three major patterns: the individualist, the hierarchical and the egalitarian. Individualist arrangements are based on the laws of the marketplace, while hierarchical relations are epitomized by government bureaucracies. Egalitarian groups are aligned in a 'border zone' on the margins of power at the political economic centre of society where the other two modes of social organization are normally located.

Egalitarian groups have a cosmology or worldview that is more or less equivalent to the 'New Ecological Paradigm' discussed by Catton and Dunlap (1978). Unbridled economic growth is frowned upon, the authority of science is questioned and our boundless faith in technology is declared unwise.

Douglas and Wildavsky's central thesis is that the perception of risk varies considerably across these three forms of social organization. Market individualists are primarily concerned with the upswing/downturn of the stock market, hierarchists with threats to domestic law and order or the international balance of power and egalitarians with the state of the environment. This leads them to conclude that the selection of risks for public attention is based less on the depth of scientific evidence or on the likelihood of danger but rather according to whose voice predominates in the evaluation and processing of information about hazardous issues.

In this view, the public perception of risk and its acceptable levels are 'collective constructs' (Douglas & Wildavsky 1982:186). No one definition of risk is inherently correct; all are biased since competing claims, each arising from different cultures, 'confer different meanings on situations, events, objects, and especially relationships' (Dake 1992: 27). Here, they are making the important point that competing definitions of what is risky are ultimately moral judgements about the proper way to organize society (Kroll-Smith *et al.* 1997: 8).

Unfortunately, at this point, Douglas and Wildavsky's cultural theory of risk slips off the rails onto spongier ground. Environmental egalitarians, they suggest, are the secular equivalents of religious sects such as the Anabaptists, the Hutterites and the Amish. Obsessed with doctrinal purity and the need for unquestioned internal loyalty, sectarians are seen as having to create an image of threatening evil on a cosmic scale. It is therefore necessary and 'functional' for environmental sectarians such as those found in Friends of the Earth to constantly identify new risks from nuclear winter to global warming. Each new crisis is chosen, they maintain, 'out of the necessity of maintaining cohesion by validating both the sect's distrust of the center and its apocalyptic expectations' (Rubin 1994: 236). This explains why they turn their back on local causes in favour of global issues so vast in scale as to warrant a sense of general doom. Pollution and other risks are wielded by these sectarian challengers as a way of holding their membership together and for attacking the establishment groups of the centre, which they oppose (Covello & Johnson 1987: x).

Risk and Culture has provoked much interest and a torrent of criticism. Much of the latter focused on the authors' claim that environmentalists mobilize for solidary rather than for purposive reasons. Rather than view environmentalism as part of a moral response to a very real societal crisis, they treat risks as merely bogeymen that serve the same purpose as certain food prohibitions among tribal peoples. Environmentalists, therefore, are not regarded as rational actors but rather as 'true believers' open to manipulation by ecological prophets such as David Brower and Edward Abbey.

Karl Dake, a member of the Douglas–Wildavsky research circle, has insisted that this criticism is overstated and that the cultural school of risk never meant to imply that perceived dangers are simply manufactured:

> People do die; plant and animal species are lost forever. Rather, the point is that world views provide powerful cultural lenses, magnifying one danger, obscuring another threat, selecting others for minimal attention or even disregard.
>
> (Dake 1992: 33)

Douglas and Wildavsky are less accommodating, however, insisting that knowledge about risk and the environment is 'not so much like a building eventually to be finished but more like an airport always under construction' (1982: 192). It is fruitless, they claim, for the social analyst to try to assess whether the risk under discussion is real or not; what matters is that the debate keeps going 'with new definitions and solutions'. Rubin (1994: 238–9) totally rejects this relativism, arguing that public policy considerations require that we know *definitively* whether risks such as those arising from global warming or ozone depletion are merely foils for the apocalyptic needs of sectarian organizations or genuine threats that must be dealt with. While Rubin's point is well taken, the ambiguity of many contemporary risks makes it difficult to achieve the certainty that he would like to see. Even if we reject Douglas and Wildavsky's absolute relativism, nevertheless, the by now widely accepted argument that they make about the subjective and imprecise nature of scientific findings militates against the infallibility of expert opinion. As a society, we still have to make social judgements about the magnitude of risk, although scientific evidence can be one helpful source of information in making these decisions.

Wilkinson (2001) has highlighted the similarities and differences between Mary Douglas and Ulrich Beck, whose 'risk society' thesis we examined in Chapter 2. Between them, he observes, 'they have provided the most detailed theoretical explanations for the social development of a new culture and politics of risk' (p. 1). Both theorists have chosen to address risk on a societal scale. Both point to the cultural relativity of risk perception and use the arguments of social constructionism. Neither is tempted to empirically investigate the prevalence of risk or the nature of risk perception. However, they differ as to the 'reality' of the risks we face. As we have seen, Beck embraces an apocalyptic vision of the future that is assured unless we engage in a new process of collaboration and social learning. By contrast, Douglas 'would cast doubt on the credibility of such an alarmist scenario and prefers to entrust herself to the professional opinion of government experts' (ibid).

Sociological perspectives on risk

Sociologists of risk generally adopt a more moderate position than that of Douglas and Wildavsky, insisting that while risk is certainly a sociocultural construct, it cannot be confined to perceptions and social constructions alone. Rather, technical risk analyses are an integral part of the social processing of risk (Renn 1992).

Dietz *et al.* (2002) observed, in preparatory work, that the main currents in the sociology of risk have followed three separate but complementary directions that are bound together by an underlying emphasis on the social context in which individual and institutional decisions about risks are made.

First, sociologists have been concerned with the question of how perceptions of risk differ across populations facing different life chances and whether the framing of choices stems primarily from power differences among social actors. Thus, Heimer (1988) points out that the residents of Love Canal saw the risks from chemical dumps differently from executives of the Hooker Chemical Company and from bureaucrats in the state government and various state agencies that deal with public health and the environment. Similarly, workers and bosses see environmental health risks in the workplace in a different light. To a certain extent, this issue overlaps the social distribution of risk, although the emphasis here is on how social location affects the perception of risk rather than on how it alters the likelihood of being exposed to hazardous conditions.

Second, sociologists of risk have proposed a model that reconceptualizes the problem of risk perception by taking into account the social context in which human perceptions are formed. Individual perception is powerfully affected by a panoply of primary influences (friends, family, co-workers) and secondary influences (public figures, mass media), which function as filters in the diffusion of information in the community. This is captured in the concept of 'personal influence' that was central in the mass communication research of the 1950s and 1960s (see Katz & Lazarsfeld 1955).

Third, risks, especially those of technological origin, have been cast as components of complex organizational systems. This is exemplified in Perrow's (1984) analysis of 'normal accidents' in which an estimated probability of failure is built right into the design of technologies with high catastrophic potential. Once implemented, however, such systems severely limit any further human ability to manipulate risks since the source of the risk is now located within the organization itself (Clarke & Short 1993).

Renn (1992) further divides the sociological approaches along two dimensions: (1) individualistic versus structural and (2) objective versus constructionist. The first dimension asks whether the approach in question maintains that the risk can be explained by individual intentions or by organizational arrangements. Objectivist concepts imply that risks and their manifestations are real, observable events while constructionist concepts claim that they are *social artefacts* fabricated by social groups or institutions. According to this taxonomy, the first two currents of risk research identified by Dietz and his colleagues tend to be individualist/constructionist while the third is structural/objective. Notable by its absence is a 'social constructionist' perspective that Renn describes as an approach that 'treats risk as social constructs that are determined by structural forces on society' (1992: 71).

Social definition of risk

Hilgartner (1992) has argued that the constructionist perspective must begin by examining the conceptual structure of social definitions of risk. Such definitions, he maintains, include three major conceptual elements: an object deemed to pose the risk, a putative (alleged) *harm* and a *linkage* alleging some causal relationship between the object and the harm.

To assume that objects are simply waiting in the world to be perceived or defined as risky is 'fundamentally unsociological' (Hilgartner 1992: 41). Rather, an initial phase of risk construction consists of isolating and targeting the object(s) that constitute(s) the primary source of a risk.

In the late 1980s, the lakeside Toronto neighbourhood in which my family and I resided was designated by the municipal public works department to receive a pair of 'sewage detention tanks', one to be installed in Kew Gardens, a multi-use community park, the other on the beach adjacent to the boardwalk. The problem, we were told, was effluent from the City's storm sewer system that flowed into Lake Ontario and made it too polluted with faecal coliform bacteria to allow swimming. According to studies conducted by an engineering firm engaged by the City, there were two primary sources from which the faecal coliform pollution originated: human faeces contained in combined sewer overflow[1] and animal excrement that had been swept along with rainwater into the storm sewers.

Our residents' association, which first learned of the project when one member came across the publication of a statutory notice buried in the pages of a local daily newspaper, at first expressed concern on the grounds of the disruption that construction would bring to the park and the beach, both of which are heavily used.

However, in the course of researching the proposal and meeting with other residents, we began to realize that, in fact, the source of the risk probably did not reside primarily in the storm-water but in effluent that was being dumped into the lake from the main sewage treatment plant located just to the west of our neighbourhood. We learned that, due to insufficient capacity, operators at this plant routinely opened the sea-wall gates just before it began to rain and released untreated or partially treated sewage into the lake at levels 10,000 times that at which the beaches were declared unsafe for swimming and closed. On one day out of three the lake currents reversed direction, sending this effluent towards our beaches. Immediately after a public meeting one night, a retired operator at the drinking-water filtration plant located at the eastern fringe of the neighbourhood told me that he routinely used to receive a telephone call from his equivalent at the sewage treatment plant advising that in advance of rain they were opening the gates and that he should raise the chlorine levels – a tip-off that the coliform pollution was migrating along the near shore area in a kind of bathtub ring pattern. We did not know it at the time but a somewhat similar situation was occurring regularly in Sydney, Australia, where the ageing sewage system, which pumps sewage out to sea, was designed to overflow into storm sewers during periods of heavy rainfall so as not to clog up already overloaded treatment tanks (Perry 1994: WS-4).

What happened here was that residents opposed to the sewage detention tanks developed an alternative definition of the 'risk object'. At public meetings, at City Hall and at a special hearing before an Environmental Assessment Advisory Committee appointed by the Provincial Minister of the Environment to consider whether to grant our request for a 'bump up' (i.e. from a routine class environmental assessment to a more formal and rigorous individual environmental assessment), we actively contested the official designation of the object deemed to be risky and presented our claim (unsuccessfully) that the main sewage treatment plant was the villain instead.

The second element in the social definition of risk involves the process of defining harm. Once again, this is not as obvious as it may seem. For example, forest fires are commonly thought to wreak a path of destruction but ecologists contend that in nature they serve a useful function in woodland renewal. Offshore oil-drilling platforms are assumed to pollute the waters surrounding them but marine biologists have found that they also spawn a whole new micro-ecology at their base. Some environmentalists in the United States have campaigned to reduce allowable levels of the trace mineral selenium, which can be added to animal rations on the basis that it leaves toxic residues, but representatives of the feed industry maintain that selenium additives are a boon to the environment because they reduce the amount of feed consumed thus saving on energy.

In each of these cases, the very definition of what harm ensues from a particular object or action is contested, sparking a variety of claims and counter-claims, despite the fact that there is mutual agreement as to the risk object (forest fires, offshore oil drilling, selenium as a feed additive). Risk claims may frequently conflict on ideational grounds; thus, a river diversion project that provides irrigation water for local farmers (a human benefit) may result in the destruction of a fragile ecosystem of fish, birds, insects, etc. (a biological harm). Similarly, road salt, deemed so vital in order to cope with the harsh winter in parts of Canada and the Northern United States, has been labelled by scientists as harmful to the ecology of the lakes, rivers and streams where it is eventually deposited. Conversely, initiatives that are declared to be of ecological benefit may result in problems for human constituencies. For example, the protection of wolves is advocated by wildlife preservationists but keenly opposed by ranchers who fear the loss of livestock crucial to their economic survival. With consensus impossible, the central basis of contestation becomes the presence or absence of harm generated by a risk object.

A third component of the social construction of risk consists of the linkages alleging some form of causation between the risk object and the potential harm. Hilgartner (1992: 42) observes that constructing these linkages is always problematic because a risk can be attributed to multiple objects. Indeed, the 'laws' of ecology encourage this since all things are regarded as being interdependent. This is further complicated by the fact that the full extent of the risk may not be known until many years later. The effects may sometimes be more immediate but it takes years for claims-makers to assemble them into a publicly acknowledged form. This has been the case with a raft of health-related ailments among military veterans of the Gulf War. Even though symptoms began soon after their return, it took some time for public reports of a 'Gulf War syndrome' to penetrate the mainstream media and to be framed in terms of toxic environmental agents in the war zone.

Much of the discourse over the construction of risk takes place on this terrain. The situation is further complicated by the existence of multiple conflicting proofs: legal, scientific, moral.

The burden of legal proof is most onerous, since it cannot leave any room for 'reasonable doubt'. The caveats that are standard in scientific studies (e.g. 'the data are suggestive but require further research') do not stand up in court. Nor usually do anecdotal or clinical evidence.[2] As environmentalists have discovered, judges are often loath to break any new ground by acting to prevent a problem before it happens. As Freudenburg (1997: 34–5) points out, the capability of the courts to deal with technological risks and disasters is especially limited by 'the need to establish clear and unambiguous liability, particularly in the presence of evidence that will remain at best probabilistic'.

Scientific proof is easier to come by, but nevertheless is a slave to statistical levels of significance. It is also notoriously fickle, its authority intact only until the next disconfirming study appears. The scientific layer of proof can be subdivided into two: a standard drawn from pure science in which action is not recommended until correlations weigh in at the 95 per cent confidence level, and a standard utilized by the medical disciplines in which action may be taken before significance is reached if the evidence points towards a serious health problem.

Collingridge and Reeve (1986) demonstrate the clash between these two versions of scientific evidence in the debate over the health effects on children of lead from vehicle exhausts. In the United States, it haunted the conflict between the EPA, which supported the removal of lead gasoline (petrol) on the basis of broad differences in blood lead levels among urban and suburban populations, and the Ethyl Corporation, a major manufacturer of lead additives, which argued that the link between blood and air levels remained statistically unproven. In the UK, difficulties arose in the early 1980s between the government-sponsored 'Lawther Report', which rejected all laboratory animal and biochemical studies as irrelevant to understanding the medical effects of lead on humans, and the report entitled *Lead or Health* by the environmental group, the Conservation Society, which argued the contrary. Moral proofs are most easily manufactured but are heavily dependent upon the mobilization of public opinion in order to make an impact.

The use of moral proofs allows the formation of attitudes or opinions about a risk issue even if the scientific or legal layers of proof indicate a degree of uncertainty or ambiguity. For example, animal rightists have never been able to conclusively prove that animals 'suffer' so they have adopted the alternative strategy of trying to demonstrate ethically that this must be the case, drawing in particular on the work of the philosopher Peter Singer. Similarly, the scientific case against the biological engineering of plants and animals is still inconclusive (no genetically altered fruits have thus far performed like the protagonist in the Roald Dahl story, *James and the Giant Peach*) but the moral case against interfering with nature is more impressive. Such moralization, however, tends to polarize positions on risk policies, making compromises more difficult (Renn 1992: 192).

Unlike the legal and the scientific, the most effective moral proofs are often those that follow a simple line of reasoning. Consider the nature of the argument presented by 'Kapox' – labelled by the South American press as the 'Tarzan of the Amazon'. Kapox, who engaged in long-distance swims through the Amazon region to publicize the state of pollution of the river and the destruction of the surrounding rainforest, did not base his appeal on a sophisticated reasoning about the need to protect biodiversity. Rather, he preached a simple, obvious, moral message: as the largest river in the world concentrating a fifth of the planet's fresh water, the Amazon deserves respect (Suzuki 1994a).

Arenas of risk construction

As powerful as Kapox's appeal may be, it is unlikely to directly influence collective risk decisions or policies. Instead, social definitions of environmental risk must be followed up by political actions designed to mitigate or control the risk that has been identified. Building on the work of Hilgartner and Bosk (1988), Renn (1992) argues that political debates about risk issues are invariably conducted within the framework of social arenas.

The term *social arenas* is a metaphor to describe the political setting in which actors direct their claims to decision-makers in hopes of influencing the policy process. Renn conceives of several different (theatre) 'stages' sharing this arena: legislative, administrative, judicial, scientific and mass media. While both traditional and unorthodox action strategies are permitted, these arenas are nevertheless regulated by an established repertoire of norms. For example, illegal direct action such as that advocated by Earth First, the American renegade environmental group, violates this protocol. The code is, in fact, a combination of formal and informal rules usually monitored and coordinated by some type of enforcement or regulatory agency such as the EPA in the United States and the Department of the Environment (D.o.E) in Britain.

The concept of the social arena combines elements from the organization–environment perspective in the field of complex organizations, Goffman's dramaturgical model of social relations and the symbolic models of politics as developed by Murray Edelman (1964, 1977) cemented together by a social constructionist compound. As formulated by Renn, it also stresses the mobilization of social resources as discussed by the McCarthy–Zald school within the resource mobilization perspective on social movements. Renn seems unaware of the parallels but the social arena concepts that he uses also echo some basic research on international environmental diplomacy, notably Haas' (1990, 1992) seminal concept of 'epistemic communities'.

While some elements of risk construction may occur in the public domain beyond their parameters, the most important action takes place in arenas that are populated by communities of specialized professionals: scientists, engineers, lawyers, medical doctors, corporate managers, political operatives, etc. (Hilgartner 1992: 52). Such technical experts are the chief constructors of risk, setting an agenda that often includes direct public input only during the latter stages of consideration. Hilgartner and Bosk (1988) note that these 'communities of operatives' often function in a symbiotic fashion, the operatives in each arena feeding the activities of operatives in the others. Environmental operatives (environmental groups, industry lobbyists and public relations personnel, political champions, environmental lawyers, journalists and bureaucrats) are notable examples of this; by virtue of their activities they both generate work for one another and raise the prominence of the environment as a source of social problems.

Within the social arena of risk, the process of defining what is acceptable and what is not is often rooted in negotiations among several or multiple organizations seeking to structure relations among themselves. Clarke (1988) illustrates this in his analysis of an office building fire in Binghampton, New York, which left a legacy of toxic chemical contamination. In this case, three governmental agencies – the state health department, the county health department and the state maintenance organization – collectively vied for suzerainty in determining how risky the situation was thought to be. In such cases, Clarke argues, the institutional assessment of risk is a claims-making activity in which corporations and agencies both compete and negotiate to set a definition of acceptable risk.

From a theatrical vantage point, social arenas of risk are populated by sundry groups of actors. Palmlund (1992) proposes the existence of six 'generic roles' in the societal evaluation of risk, each of which carries its own dramatic label: risk bearers, risk bearers' advocates, risk generators, risk researchers, risk arbiters and risk informers.

Risk bearers are victims who bear the direct costs of living and working in hazardous settings. In the past, those who are impacted most have rarely asserted themselves and have therefore remained on the margins of risk arenas. More recently, however, as can be seen in the rise of the environmental justice movement, risk bearers have become empowered and must increasingly be regarded as notable players. *Risk bearers' advocates* ascend the public stage to fight for the rights of victims. Examples include consumer organizations such as those headed by Ralph Nader and Jeremy Rifkin, health organizations, labour unions and congressional/parliamentary champions. They are depicted as protagonists or heroes. *Risk generators* – utilities, forestry companies, multinational chemical/ pharmaceutical companies, etc. – are labelled as antagonists or villains since they are said by advocates to be the primary source of the risk. *Risk researchers*, notably scientists in universities, government laboratories and publicly-funded agencies are portrayed as 'helpers' attempting to gather evidence on why, how and under what circumstances an object or activity is risk-laden, who is exposed to the risk and when the risk may be regarded as 'acceptable'. On occasion, however, risk researchers have become identified with risk generators, particularly if their findings support the latter's position. *Risk arbiters* (mediators, the courts, Congress/ Parliament, regulatory agencies) ideally stand off-stage, seeking to determine in a neutral fashion the extent to which risk should be accepted or how it should be limited or prevented and what compensation should be given to those who have suffered harm from a situation judged to be hazardous. In reality, risk arbiters are rarely as neutral as they should be; instead, they frequently tend to side with risk generators. Finally, *risk informers*, primarily the mass media, take the role of a 'chorus' or messengers, placing issues on the public agenda and scrutinizing the action.

Renn (1992) suggests a hybrid of several of these roles: the *issue amplifiers* who observe actions on stage, communicate with the principal actors, interpret their findings and report them to the audience. Environmental popularizers such as Paul

Ehrlich, Barry Commoner, Bill McKibben, Jeremy Rifkin and Jonathan Porritt are prime examples of this.

Hilgartner and Bosk (1988) depict the interaction among different arenas of public discourse as characterized by several key features. First, these multiple arenas are connected by a complex set of linkages both social and organizational. As a result, activities in each arena thoroughly propagate throughout the others. Second, one finds a huge number of 'feedback loops' that either amplify or dampen the attention given problems in public arenas. Consequently, you find a relatively small number of successful social problems that occupy much of the space in most of the arenas at the same time. This synergistic pattern is typical of policy-making on matters relating to risk and the environment.

In their study of 228 Washington-based 'risk professionals', Dietz and Rycroft (1987) found a policy community with a dense network of communication that stretched across environmental groups, think-tanks, universities, law and consulting firms, corporations and trade associations, the EPA and other executive agencies. Environmental organizations were especially active in outreach activities including contacts with corporations and trade associations with whom 85 per cent of respondents communicated in a typical month. Similarly, the personnel flows across organizations, another component of the exchange network, was substantive, although working for an environmental group led to a low probability of finding employment with one of the other groups.

Dietz and Rycroft depict the environmental risk policy system as a hybrid in the sense that it has a strong base in science but, at the same time, is driven by the ideological conflict between environmentalists and corporate and govern-mental participants. This creates a measure of volatility insomuch as science is the cornerstone of the system yet many key decisions are resolvable only in political terms. Nevertheless, the picture that emerges from this survey study is one of a policy community that is permeable but closely linked and oriented towards a shared discourse on issues relating to environmental risk. Among other things, this means that any approach to risk that attempts to privilege socio-cultural facts over material ones will probably be considered off target and therefore inappropriate for inclusion on the shared agenda of risk professionals (Dietz & Rycroft 1987: 114).

Power and the social construction of environmental risk

Freudenburg and Pastor (1992) have observed that the social constructionist approach to risk is well positioned to discuss risk construction in the context of power. In a similar fashion, Clarke and Short (1993) note that constructionist arguments – in contrast to those anchored in psychology and economics – tend to focus on how power works in framing terms of debate about risk.

Both sets of authors share the belief that this relationship is especially important because official viewpoints, with their significantly greater access to the mass media, strongly suggest that public fears regarding technical risks are clearly irrational; that is, claims about public irrationality are in themselves ways of framing risk issues. By implication, policy formulations that originate with the community of risk professionals (see the previous section) are presented as rational, objective assessments of what is considered safe and what is not. If this view is accepted, then the central task is said to be educating the public to realize that they are over-reacting and that the risk attached to nuclear power, herbicides, bioengineered organisms, etc., are not really the hazards that they appear to be. In order to allay public fears, risk analysts develop quantitative measures through which to compare the risks inherent in different policy choices and their relative costs and benefits (Nelkin 1989: 99).

This is not to imply that the people are always right and the knowledge of the experts invariably 'brittle' (Wynne 1992). Rather, a social constructionist perspective would argue that each represents a competing frame but the dominant rationality that comes from the risk establishment is superimposed over the popular frame due to a power differential. Thus, Wynne (1992: 286) demonstrates in the case of a public controversy over the herbicide 2,4,5-T in the United Kingdom that the firsthand empirical knowledge of farm and forestry workers was directly relevant to an objective risk analysis. However, scientists flatly refused to consider this knowledge as legitimate, thereby denigrating and threatening the social identity of the local citizens.

Nowhere is this differential more evident than at public information meetings or hearings that are routinely stage-managed by risk generators and arbiters. At the public meetings concerning the building of the sewage detention tanks described earlier in this chapter, members of the public works department, local politicians (who strongly supported the project) and representatives of the private engineering firm who had recommended the building of the tanks all sat together on the elevated stage of the auditorium whose perimeters were adorned with charts, blown-up photographs and other 'props'. We citizens were restricted to a single question with no follow-up. Those who queried the suitability of the project were alternately bullied and patronized. On contentious issues the presenters did not hesitate to introduce a ream of previously unseen statistical evidence that we had no way of confirming or denying without days or weeks of further research.

Richardson *et al.* (1993) observed many of the same structural elements in the conduct of a set of environmental public hearings in 1984 on the proposed building of a bleached kraft pulp mill in northern Alberta, Canada.[3] The members of the Alpac Environmental Impact Assessment Review Board, who were conducting the hearing, sat at a table facing the public on a stage. At one or several tables to the direct right of the Board were the representatives of Alberta-Pacific Forest Industries (Alpac), the company that sought to build the mill, their technical experts and their

Box 7.1   Toxic talk

According to Kaminstein (1988), the primary tool that scientific experts associated with the EPA and the Centers for Disease Control use to stifle citizen initiatives is toxic talk – talk that stifles discussion and smothers public concern. The rhetoric of containment has multiple elements.

First, as was the case with the detention tank meetings in Toronto, residents of Pitman, New Jersey were bombarded with technical information. At one meeting, EPA officials distributed documents totalling 44 pages. Those in attendance were expected to assimilate an array of data, charts, graphs, tables and a slide show in rapid succession. At the same time, the facts that residents wanted were never available, and no explanation or interpretation was given as to the information that the consultant scientists presented.

The physical setting of the meeting room was also very similar to that experienced by those attending the detention tank sessions. At the front of the room was a large dais raised about two feet, a long table and nine large, high-backed chairs on which the scientists sat, creating a physical and psychological distance from the audience. Various dramatic props, for example, an enlarged photograph of an air-monitoring vehicle that looked like a recreational camper, were employed as rhetorical devices to pacify the residents and enhance the power of those in charge of the meeting.

The factual presentation style used by EPA officials and scientists was abstract, impersonal and technical, thus creating an impression of professional neutrality. It was the activist residents who became angry and confrontational, allowing officials to dismiss them as 'emotional'. Questions that dealt with the geology and hydrology of the area, future tests and plans for the clean-up were addressed but those that dealt with health risks were avoided or deflected. Officials and scientists used language in their presentations that was technical, ambiguous and intellectual, making it impossible for any meaningful dialogue to develop between experts and residents over the nature and magnitude of the risks faced by the community of Pitman.

Toxic talk techniques such as this are strategically successful if ethically reprehensible. It allows scientific experts and government officials to direct the discussion, set the risk agenda and discourage future citizen participation. Popular concerns and risk frames are subordinated to those that are preferred by the powerful in society. As Kaminstein (1988: 10) notes, these kinds of exclusionary devices permit agencies such as the EPA to legally fulfil their mandate to hold public meetings while at the same time leaving residents feeling that they are fighting a losing battle just to be heard.

lawyer. Numerous Alpac consultants were scattered throughout the proceedings. Presenters were required to speak into microphones through which their words were recorded.

Kaminstein (1988) argues that embodied in the public presentation of scientific information at meetings concerning the health and safety aspects of toxic waste dumps is a *rhetoric of containment* that restricts discussion, avoids tough questions and pursues its own agenda. Drawing on three years of observation meetings held to inform residents of Pitman, New Jersey, about the steps being taken to clean up the Lipari landfill, the site of one of the worst dumps in the United States, Kaminstein concludes that residents were not so much informed or persuaded as controlled and defeated (see Box 7.1).

That is not to say that members of the public never attempt to assert themselves in settings such as these. For example, in the Alberta case, some participants fought to wrest control from regulators over the scope of the review, the venues and over definitions of legitimacy, as well as attempting to subvert the dominant discourse that was imposed by the pro-development forces (Richardson *et al.* 1993: 47). However, the constraints of the hearing process normally make effective citizen participation difficult, especially since the situation is structured so as to prevent public argument and reinforce the power of institutions.

Institutional risk analysts and regulators also exercise power on a broader plane. Structurally, they control the official risk agenda, acting as gatekeepers who are well placed to determine which issues are included or excluded from public discourse. For example, in the 1980s, imbued with the deregulatory climate within the Reagan administration (supported by senior EPA managers), Congress fatally slashed the budget of the Office of Noise Abatement and Control (ONAC) thereby also dooming most state and local noise abatement programmes (Shapiro 1993). Despite the continued risk posed by noise pollution to human health and environmental aesthetics, the issue stalled for lack of government action. In such circumstances, the risk itself does not diminish (in the case of noise pollution it in fact increased) but the risk establishment is able to manipulate its progress on the action agenda.

Freudenburg and Pastor (1992: 403) note that the social constructionist approach to technological risks would do well to look at other variables that sociologists have previously found to be associated with power. Gender may be significant here, insomuch as the scientific experts and bureaucratic officials who practise the rhetoric of containment are usually men while local citizen groups are disproportionately composed of women, many of whom lack power and authority in public life. Similarly, members of racial and ethnic minorities are routinely dismissed and discredited by the risk establishment, an experience that has led to the blossoming of the environmental justice movement. The relationship between power, inequality and the social construction is equally evident in communities that have been marginalized by positions of economic, geographic or social isolation (Blowers *et al.* 1991).

Risk construction in cross-national perspective

Finally, risk construction varies cross-nationally according to a number of different factors: the organization of political and administrative structures, historical traditions and cultural beliefs. Within the field of risk analysis, a classic comparative study is Sheila Jasanoff's (1986) report entitled *Risk Management and Political Culture*. Drawing on case studies of national programmes for controlling carcinogens in several European countries, Canada and the United States, she concludes that cultural factors strongly influence goals and priorities in risk management. In Germany, the favoured approach has been to delegate resolution of all risk-related issues to technical experts. Jasanoff does not discuss it, but even where a risk subject is strongly contested, technical rationality is applied in the form of a 'technology assessment' that includes representatives from government, industry and social movements (see Bora & Dobert 1992). In Britain and Canada, risks are examined through a mixed scientific and administrative process but scientific uncertainties are not always publicly broadcast. By contrast, in the United States risk determination has a much more public face surfacing in a wide variety of administrative and scientific fora. While this can produce greater analytical rigour and more democratic and informed public participation, it can also lead to more polarization and conflict and thus to political stalemate.

Using the comparative method suggested by Jasanoff, Harrison and Hoberg (1994) compared government regulation in Canada and the United States of seven controversial substances suspected of causing cancer in humans: the pesticides alar and alachlor, urea-formaldehyde foam insulation, radon gas, dioxin, saccharin and asbestos. Each country's approach was weighed according to five criteria for effectiveness: stringency and timeliness of the regulatory decision, balancing of risks and benefits by decision-makers, opportunities for public participation and the interpretation of science in regulatory decision-making.

As had Jasanoff, the researchers found that there were two contrasting regulatory styles. In each case:

> there was more open conflict over risks in the United States than Canada, with interest groups, the media, legislators and the courts playing a much more important role south of the border. The regulatory process in Canada tended to be closed, informal, and consensual, in comparison with the open, legalistic, and adversarial style of the US.
>
> (Harrison & Hoberg 1994: 168)

Both styles are said to have risks and benefits. The Canadian system is more conducive to scientific caution and formal democratic control but it lacks accountability, making it easier for political decisions to be cloaked in scientific arguments. The American system is more open but also more conflictual and vulnerable to interest group pressures and, as a result, less dependent upon scientific expertise.

This comparative research provides further evidence that risk determination and assessment are socially constructed. National political structures and styles can be seen to have as much to do with deciding which environmental conditions will be judged to be risky and actionable as the nature of the scientific claim itself. Consequently, fundamentally sound environmental claims may be deflected or stalled, either due to collusion between regulators and scientists or because of political pressure from interest groups, either within or opposed to the environmentalist perspective.

Conclusion

In the social sciences, two conflicting ways of looking at risk have clashed over the past three decades. On the one hand, you have an engineering-style approach that treats risks as 'objectively' determinable, if only you have the right instruments and information. By contract, a 'sociological' perspective on risk insists that risk is as much about the social as the physical; that is, perceptions of risk differ by class, gender, race, ethnicity, ideology and past experience. Whether an object or situation is judged to be 'risky' differs depending on whether you apply legal, scientific or moral proofs. In the literature on risk, an important concept is that of the 'social arena', which refers to the political setting in which claims about risk are made and contested. The social constructionist approach is especially valuable here because it focuses on how power works in framing the terms of debate about risk.

Notes

1   For many years, human sewage from many local households mixed together with storm water in the same pipe. There has since been a vigorous sewage separation programme, but some homes and businesses still discharge sewage into the storm-water system.
2   One exception to this is a 1984 decision in the *Ferebee v. Chevron Chemical Co.* case in the United States that allowed the jury to rely on the testimony of individual physicians in the absence of ironclad epidemiological evidence concerning injury by exposure to pesticides (see Cronor 1993).
3   As it happens, the Review Board recommended that the mill should not be built unless further studies indicated that it would not pose a serious hazard to biological life in the river and for downstream users along the Peace-Athabasca river system. Nine months after it agreed to abide by these findings, the Alberta government overturned its own decision and decided to allow Alpac to proceed.

Further reading

Douglas, M.A. and Wildavsky, A. (1982) *Risk and Culture: An Essay on the Selection of Technological and Environmental Dangers*, Berkeley, CA: University of California Press.
Freudenburg, W.R. and Gramling, R.B. (2011) *Catastrophe in the Making: The Engineering of Katrina and the Disasters of Tomorrow*, Washington, DC: Island Press.

Lash, S., Szersynski, B. and Wynne, B. (eds) (1996) *Risk, Environment and Modernity: Towards a New Ecology*, London: Sage.

Perrow, C. (1999) *Normal Accidents: Living With High Risk Technologies*. Princeton, NJ: Princeton University Press.

Online resource

Society for Risk Analysis. Available HTTP: http://sra.org

8 Biodiversity loss

The successful 'career' of a global environmental problem

Along with global warming, the conservation of biodiversity was one of the two major issues at the June, 1992 United Nations Conference on Environment and Development (UNCED) in Rio de Janeiro. It was called the 'hottest' environmental topic of 1993 (Mannion 1993), with a burgeoning academic and popular literature devoted to exploring its parameters. Six years later, Valiverronen (1999: 404) characterized it as 'the latest "big" environmental issue, comparable to acid rain, ozone depletion and climate change'. Yet, 20 years before, the term biodiversity was unknown and it was not to be found in any compendium of threats to the environment. The skyrocketing career of biological diversity loss is a good illustration of how a 'transnational epistemic community' (see Chapter 6) can assemble, present and successfully contest a global environmental problem.

As a concept, biodiversity is multi-layered with various levels of meaning (Udall 1991: 82). Officially, it has been defined as 'the variability among living organisms from all sources, including *inter alia*, terrestrial, marine and other aquatic ecosystems and the ecological complexes of which they are part' (Tolba & El-Kholy 1992). More simply, it is an umbrella term for nature's variety – ecosystems, species and genes (Environmental Conservation 1993: 277). Lorimer (2006: 540) argues that our view of biodiversity has been skewed by an 'objective understanding' as outlined at the 1992 Earth Summit according to which 'biodiversity is constituted by a set of objects and processes revealed to us by an all-seeing, disembodied natural science'.

Biodiversity is generally acknowledged to exist at three distinct levels: ecosystem diversity, species diversity and genetic diversity.

Ecosystem diversity refers to the variety of habitats that host living organisms in a particular geographic region. This variety is said to be shrinking in the face of accelerating economic development. Udall (1991: 83) uses the metaphor of a ripe pumpkin that has been hollowed out to describe the damage to our ecosystems, which has been inflicted by trapping, ploughing, logging, damming, poisoning and other forms of human intrusion. With the rapid pace of development, land ecosystems are described as increasingly taking the form of 'habitat islands'; for example, a patch of tropical forest surrounded by croplands (Franck & Brownstone 1992: 37).

Species diversity refers to the variety of species that are found in an ecosystem. While there have been notable episodes of species extinction in the past, the scale of loss today is judged to be unprecedented in the history of humankind (Lovejoy 1986: 16). Much of this is attributable to loss of ecosystem diversity; as a broad general rule, reducing the size of a habitat by 90 per cent will reduce the number of species that can be supported in the long run by 50 per cent (Tolba & El-Kholy 1992: 186).

Genetic diversity refers to the range of genetic information coded in the DNA of a single population species. Biologists value genetic diversity because it is seen as the basis for permitting organisms to adapt to environmental change. For example, in agriculture, wild strains of plants are valued because they often contain genes that are vital in fighting off pests or disease, unlike domesticated 'monocultures' that are much more vulnerable. In the animal world, inbreeding among a population stranded by habitat loss or commercial exploitation leads to an inability to survive in the long term; for example, this is the situation of the grizzly bears in Yellowstone Park in the American West (Udall 1991: 82).

When all three levels are viewed together, biodiversity loss appears to be a newly minted environmental problem. However, as Barton (1992: 773) has observed, there have long been a variety of treaties governing individual elements such as the international trade in endangered species, regional conservation and the conservation of particular species. For example, in 1911, a major international agreement, the Convention for the Protection and Preservation of Fur Seals, had been signed; the Migratory Birds Convention, signed in 1917 by the United States and Canada, was also a key piece of legislation in the campaign during the first part of this century to save North American birds.

Contextual factors

There were three major developments that set the stage for the rise of biodiversity loss as a major environmental problem in the 1980s and 1990s.

First, the growing economic importance of biotechnology meant that a greater financial value was increasingly being placed on genetic resources, a value that was recognized through intellectual property rights (Barton 1992: 773). Of special importance here was a landmark decision by the US Supreme Court (*Diamond v. Chakrabarty*) that allowed for the first time the patenting of a genetically engineered microbe, in this case an oil-eating bacterium developed by a General Electric research scientist named Ananda Chakrabarty. Also of significance was the passage a decade earlier of the US Plant Variety Protection Act (PVPA) that set up a patent-like system to govern the seed industry under the auspices of the US Department of Agriculture rather than under the more rigorous requirements of the US Patent Office. These events were significant for two interrelated reasons.

By raising the monetary stakes involved in the development of genetic resources, a conflict was fanned between the developed nations who wished to ensure open access to plant and animal genes and the less developed nations in which the bulk of these genetic materials were to be actually found. The latter began to see the genetic prospecting of the multinational pharmaceutical and chemical companies headquartered in Northern nations as a form of 'plundering' for which compensation should be paid.

At the same time, genetic diversity also became an international development issue due to the entry of several well-known rural activists (Cary Fowler, Pat Roy Mooney) to the debate over plant patenting. Fowler, a farmer from North Carolina, had worked with food activists Frances Moore Lappé and Joe Collins on the national bestselling book *Food First*, an indictment of the world food system. Fowler became a one-person lobby opposing changes to the seed patent laws. In the 1979 debate over a proposal to amend the PVPA so as to add six 'soup vegetables' theretofore excluded from the act, Fowler

> turned his mailing list loose on Congress, went to the Press, wrote articles about the issue, and traveled around the country alerting other groups to the 'seed patenting' issue. Fowler rallied scientists and church interests and wrote to the Secretary of Agriculture, Bob Bergland, urging him to consider the impact of rising seed costs on small farmers.
>
> (Doyle 1985: 67–8)

Mooney, a Canadian from the province of Manitoba, helped to internationalize the seed issue both by his participation in a network of activist scholars working on Third World issues and through his widely circulated paperback book, *Seeds of the Earth*, published in 1979 for the Canadian Council for International Cooperation and the International Coalition for Development Action.

Second, the emergence of conservation biology in the late 1970s as an academic specialty provided a nesting spot for research on biodiversity. Conservation biology is an applied science that studies biodiversity and the dynamics of extinction. It differs from other natural resource fields such as wildlife management, fisheries and forestry by accenting ecology over economics (Grumbine 1992: 29). The role of the conservation biologist is to provide 'the intellectual and technological tools that will anticipate, prevent, minimize and/or repair ecological damage' (Soulé & Kohm 1989: 1). Conservation biology is thus a 'crisis discipline'[1] that draws its content and method from a broad range of fields within and outside of the biological sciences.

Conservation biology was formally recognized as a discipline in 1985 with the creation of the Society for Conservation Biology (SCB). Within three years, the membership of the Society had swollen to nearly 2,000 members (Tangley 1988: 444). SCB is significant because it has provided a central forum for the communication of knowledge about conservation and biological diversity, especially through its journal, *Conservation Biology*.

Another critical node in the development of the discipline was the establishment of the Center for Conservation Biology (CCB) in the Department of Biological Sciences at Stanford University in California. The Center's main activities are basic and applied research, education and the application of conservation biology principles to genetic resources, species, populations, habitats and ecosystems. CCB consults not only within the United States but also internationally, especially in Latin America (Franck & Brownstone 1992: 66), providing yet another link between research on biological diversity and the international development scene.

By the late 1980s, conservation biology had begun to develop rapidly at institutions of higher learning. A pioneering textbook, *Conservation Biology: Science of Scarcity and Diversity*, had been adopted by classes at 37 US colleges and universities as well as overseas (Tangley 1988: 444). In 1985, the first conservation biology course was taught at the University of California at Berkeley with an emphasis on the biological foundations for conservation (Millar & Ford 1988: 456). While research funding was still modest, partly because of a perception that conservation biology was a 'soft' discipline, advocates of biological diversity as an environmental problem nevertheless had an increasingly powerful academic medium for spreading their message and for building a constituency.[2]

Third, a legal and organizational infrastructure was being assembled in the 1970s within the United Nations and other NGOs[3] dealing with various elements of the biodiversity loss problem.

In 1971, the *Convention on Wetlands of International Importance Especially as Waterfowl Habitat* was agreed upon with the dual purpose of designating environmentally sensitive areas for migratory waterfowl and facilitating trans-border cooperation among countries situated along their travel routes. This agreement was staffed by a secretariat provided by the International Union for Conservation of Nature (IUCN).

The *Convention Concerning the Protection of the World Cultural and Natural Heritage* (held in Paris in 1972), prepared under UNESCO (United Nations Economic, Social and Cultural Organization) supervision, established exceptional World Cultural Sites such as Serengeti National Park in Tanzania, the Queensland Rainforests in Australia and Great Smokies National Park in the United States, some of which rated quite highly in biological diversity. The agreement established a world heritage fund to assist nations that may have difficulty in paying for the protection of these unique sites. It was signed by 150 countries. However, this treaty is extremely limited in scope and has had minimal success both in slowing the rate of species loss on a global scale and in assuring the protection of designated sites (Spray & McGlothlin 2003b: 154).

In 1973, the *Convention on International Trade in Endangered Species of Wild Fauna and Flora* (CITES) was proclaimed in Washington with a Secretariat staffed by the United Nations Environment Programme (UNEP) located in Lausanne, Switzerland. This convention established lists of endangered species for which international trade is to be controlled via permit systems. CITES was limited, however, insofar as it was directed at individual species rather than the habitats in which they resided. Furthermore, it is primarily a trade agreement that does not guarantee protective status or conservation programmes within the states in which vulnerable species reside (Spray & McGlothlin 2003b: 155). Finally, the monitoring and enforcement of CITES has been marred by a series of 'exceptions', for example, the 'tourist souvenir exception' (allows rare specimens to be imported as personal or household effects) and the 'trans-shipment exception' (permits specimens passing through a third country to avoid regulations of the convention). These exceptions have been used to smuggle protected species under the pretense that the exceptions apply (Louka 2002: 116–17).

Finally, the *Convention on Conservation of Migratory Species of Wild Animals (CMS)*, also known as the *Bonn Convention*, provides a framework for international cooperation among states that hosted animals whose travels regularly take them across national boundaries. A central aim of this convention is to coordinate research, management and conservation resources such as habitat protection and hunting regulation affecting migratory species.

These international legal agreements were supplemented by a number of regional measures, for example, conventions on the conservation of nature, natural resources, wildlife and natural habitats pertaining to the South Pacific (1976), Africa (1968) and Europe (1976), and by the designation of Biosphere Reserves under a UNESCO programme. Taken as a whole, such measures were not only useful in their own right as a means of fostering, if not enforcing, useful cooperation among nations in conserving biological diversity, but they also put into place a global system upon which more far-reaching and stringent international legislation to conserve biological diversity could be modelled. Furthermore, they established epistemic networks of research, communication and coordination that were vital in moving biodiversity along to its status today as a major environmental problem.

Assembling the claim

In contrast to those global environmental problems that involve damage by pollutants to the atmosphere (or stratosphere) – global warming, ozone depletion, acid rain – the threatened loss of biological diversity has been less dependent on the discovery of an alteration in nature; for example, the ozone 'hole' over the Antarctic or 'forest die-back' in the Black Forest. Rather, it has developed in the context of a steady outpouring of studies that have cumulatively raised the alarm.

Taken as a whole, these studies have often lacked precision, with the result that the projected number of extinctions that might be expected has varied not only widely but wildly (Brown 1986: 448). Estimates have frequently been made in terms of rates, a device that both implies a greater accuracy than is possible given current knowledge, leading to some questionable figures. Most notably, the 'one extinction per minute' rate used by some authors is equivalent to 525,600 extinctions per year, an unlikely or impossible total about ten times the number usually cited (Lovejoy 1986: 14). At the lower end, USAID (United States Agency for International Development) currently claims that 1,000 species per year are becoming extinct (http//www.usaid.gov/our_work/environment/biodiversity). One explanation for the wide disparity here is the data source. High estimates of extinction such as that contained in Greenpeace advertisements (50,000–100,000 species becoming extinct every year) are based on mathematical models. By contrast, low estimates – IUCN-WCU puts this at one bird and mammal species per year – represent *observed* rates (Budiansky 1995: 165; Kellow 2007: 18–19).

Furthermore, the enormity of the problem has meant that reliable information is difficult, if not impossible, to assemble. So little is actually known about how species interact in ecosystems, about how they depend upon each other and about how they recover from episodes of disturbance that 'actions required now to avoid future disasters must be undertaken without sufficient knowledge to make considered choices' (Norton 1986: 11).

Most current methodologies for the assessment of biodiversity use either of two methods: the measurement of species and the identification of genetic diversity. The former is inadequate insofar as it is not always the appropriate unit of measurement (use of phyla and families may be more accurate), it is not necessarily the best way of locating diverse ecosystems and it doesn't provide for changes in species and habitats over time. Identification of genetic diversity is even more difficult, insofar as it is expensive, requires trained personnel capable of using sophisticated laboratory techniques, and produces difficult-to-interpret results (Louka 2002: 124).

Finally, some scientists have questioned whether existing efforts to quantify biodiversity loss rates are flawed because they incorrectly assume that extinctions are 'random'. Thus, Raffaelli (2004) has argued that in the real world most extinction events are nonrandom; that is, some species are more likely to go extinct than others. Such nonrandom extinctions may have greater consequences for species loss than those predicted on the basis of studies in which extinctions are assumed to occur randomly (p. 1142).

In the face of this scientific uncertainty, those who have promoted biodiversity as an environmental problem have fallen back on the 'precautionary principle' (see Chapter 6), which recommends that the wisest course is simply to avoid actions that needlessly reduce biological diversity (Tolba & El-Kholy 1992: 197).

How, then, were conservation biologists and other claims-makers able to elevate biodiversity loss to the status of a notable environmental problem, given a relative lack of authoritative research data on the subject?

Wilson (1986: v) believes that the rising interest during the 1980s among scientists and portions of the public in matters related to biodiversity and international conservation can be ascribed to two more or less independent developments.

The first was the convergence of data from three different areas of research – forestation, species extinction and tropical biology – such that global biodiversity problems were brought into sharper focus and thought to warrant broader public exposure. This critical mass of data was not sufficient to build an airtight case for immediate worldwide action but it raised the profile of biodiversity to a level sufficient to provoke a stream of academic conferences, political hearings and public forums.

The largest of these was the National Forum on Biodiversity held in Washington, DC on 21–24 September, 1986 under the auspices of the National Academy of Sciences and the Smithsonian Institution. This forum featured 60 leading biologists, economists, agronomists, philosophers and international development experts. The lectures and panels were regularly attended by hundreds of people and the final evening's panel was teleconferenced to an estimated audience of 5,000 to 10,000 at over 100 universities and colleges in the United States and Canada. It was at this conference, Wilson (1986: vi) recalls, that the term 'biodiversity' was first introduced by the organizer Dr. Walter G. Rosen, a programme officer from the Commission on Life Sciences, National Research Council/National Academy of Sciences. It is also worth noting that in spite of Wilson's protests that the term biodiversity was too 'catchy' and 'lacks dignity', Rosen and the other Academy staff members persisted on the grounds that the term was simpler and more distinctive than biotic diversity or biological diversity, and therefore the public would remember it more easily (Wilson 1994: 359).

The second development was the growing awareness of the close link between the conservation of biodiversity and economic development, especially in Southern nations. This elevated biodiversity loss from a scientific environmental problem to a wider status as a sociopolitical problem. In the industrial nations of the North, destruction of tropical rainforests and other Southern habitats was decried on the basis that it threatened a vast untapped reservoir of species that could potentially prove useful in providing new foods, medical treatments and other products. At the same time, in the countries of the South, biodiversity loss was feared for its impact on local farmers and others whose livelihoods depend on the maintenance of traditional ecosystems. In time, these two objectives were to come into direct conflict, but initially they acted in concert so as to reframe biodiversity loss as a 'development' problem of considerable importance. According to USAID, the net economic benefits of biodiversity in 2005 were estimated to total at least US$3 trillion per year, or 11 per cent of the annual world economic output (www.usaid. gov/our_work/environment/biodiversity).

This integration of conservation and development found a significant funding source in the US Agency for International Development (USAID), which expanded into the area in the 1980s by mandate from Congress. In addition to sponsoring individual projects and conferences in lower income countries, USAID administers a sizeable peer-reviewed research grant programme. The centrepiece of the latter is the Conservation of Biological diversity project. This has two main components: cooperative funding of National Science Foundation (NSF) grants for research that contributes to the conservation of biodiversity in developing countries, and core funding for the Biodiversity Support Program, a consortium formed by the World Wildlife Fund, The Nature Conservancy and the World Resources Institute. USAID projects have been carried out in Latin America, the Caribbean, Africa and North Africa as well as Europe and Asia (Alpert 1993: 630). By the early 1990s, the agency was investing about US$100 million a year in biodiversity programmes around the world (Angier 1994). Today, it supports conservation activities in 64 countries.

The assembly of biodiversity loss as an environmental problem benefited greatly from the participation of several well-known scientific news entrepreneurs or champions who were extremely active in promoting it both within and beyond the parameters of science.

The Ehrlichs, Paul and Anne, had already achieved a measure of fame (as well as notoriety) in the late 1960s and early 1970s for their campaign to make overpopulation the centrepiece of the environmental crisis. Subsequently, the two biologists turned their attention to the problem of biodiversity loss. In 1986, they founded the Center for Conservation Biology at Stanford University with Paul Ehrlich as president. In 1981, the Ehrlichs published *Extinction*, one of several high-profile books that appeared on the topic of endangered species and biodiversity around this time. Here they infused the biodiversity problem with a moral dimension using the 'Noah Principle' to claim that the foremost argument for the preservation of all non-human species is the religious belief 'that our fellow passengers on Spaceship Earth . . . *have a right to exist*'.

A second major champion of the conservation of biodiversity is the celebrated Harvard entomologist Edward Wilson.[4] Widely known as one of the founders of the field of 'sociobiology', Wilson is also a 1978 Pulitzer Prize-winning author whose bestselling book, *The Diversity of Life*, was carried as a selection by book clubs across the United States and Canada. In his autobiography, Wilson reports he was 'tipped into active engagement by the example of my friend, Peter Raven', who had been writing, lecturing and debating the issue of mass extinction since the late 1970s. Among his contributions, Wilson edited a key collection of articles arising out of the 1986 National Forum under the title *Biodiversity* (1988); this became one of the bestselling books in the history of the National Academy Press (Wilson 1994: 358).

Other key figures in assembling the problem of biodiversity loss were Peter Raven, director of the Missouri Botanical Garden; Norman Myers, a well-known international conservationist who published the book *The Sinking Ark* in 1979 and Michael Soulé, founder and popularizer of the discipline of conservation biology.

Longtime friends who had similar interests, moved in the same circles and often did fieldwork in the same areas (Mazur & Lee 1993: 703), Ehrlich (Paul), Raven and Wilson were involved in many of the same endeavours to promote biodiversity loss as a global environmental problem, and they organized and chaired panels at the 1986 forum in Washington. In 1989, Raven and Wilson gave expert testimony before the US Senate Subcommittee on Environmental Pollution. Wilson and Ehrlich were contributors to the special biodiversity issue of *Science* in August 1991, and all three scientists were founding members of the Club of Earth, an activist group formed to bring scientific attention more quickly to important but neglected environmental problems (Brown 1986). Mazur and Lee observe of this trio:

> Their research productivity, their eminence and their social and institutional contacts gave them strong voices within the scientific establishment and good access to Federal and private sources of funding, which supported both their scientific and policy efforts.
>
> (1993: 703)

Wilson (1994: 357–8) refers to a 'loose confederation of senior biologists that I jokingly call the "rainforest mafia"', whose members, besides Raven and himself, included Jared Diamond, who a decade later went on to become the bestselling author of several influential books on societal collapse (see Chapter 1), Thomas Eisner, Daniel Janzen, Thomas Lovejoy and Norman Myers. All of these scientists were instrumental in advancing claims about the importance of biodiversity loss.

Presenting the claim

In presenting biodiversity as an environmental claim and keeping it on the public agenda, proponents face three formidable problems (McNeely 1992: 25).

First, unlike some other environmental problems such as toxic dumps or oil spills at sea, there is no easily identifiable opponent against which public opinion can be galvanized. Instead, the root causes of biodiversity loss are found in basic economic, demographic and political trends including the relentless human demand for commodities from the tropics, runaway population growth and the escalating debt burdens of Southern nations (McNeely *et al.* 1990b).

Second, the loss of biodiversity has no immediate impact on human lifestyles in the First World nations where the resources that could be applied to acting upon the problem are concentrated. Indeed, with the exception of a small number of 'charismatic megafauna' – whales, gorillas, whooping cranes, bald eagles – most threatened organisms consist of creatures such as fungi, insects and bacteria that most people would not hesitate to step on (Mann & Plummer 1992: 49). This problem is even more exaggerated at the system level where, as Noss (1990) has sardonically observed, 'you can't hug a "biogeochemical" cycle'. As Mazur and Lee note,

The plights of these [charismatic] animals became salient through popular books, television documentaries such as those produced by the National Geographic Society, and news coverage of a few effective spokespersons including Jacques Cousteau, Brigitte Bardot, Roger Tory Peterson and Jane Goodall, who usually specialized in a single type of animal.

(1993: 701)

Third, the collective benefits of taking action are notably imprecise. At best, conservationists can speculate that somewhere in the vanishing rainforest lies the cure for cancer or AIDS but there are no ironclad guarantees. By contrast, the costs of implementation are more apparent and onerous especially on the domestic front in developed nations. As a result, public attention often begins to lag when the visible costs seem to outweigh the immediate benefits. Public support can be further eroded by high-profile controversies such as that which occurred in the United States in the late 1980s and early 1990s over the fate of the Northern spotted owl.[5]

Claims-makers have addressed these difficulties by adopting a 'rhetoric of loss' (Ibarra & Kitsuse 1993). Public statements by conservation biologists and other policy entrepreneurs stress that we are 'at a crossroads in the history of human civilization' (McNeely *et al.* 1990b: 40). Failure to act decisively is equated with turning down the road to chaos or driving a business into liquidation. Many of these metaphors are borrowed from the rhetoric of another environmental problem – overpopulation. Once again, we are depicted as rapidly approaching the 'limits of growth', thereby running the risk of surpassing the 'carrying capacity' of the planet. Lester Brown (see Chapter 1) uses the rhetorical motif of a 'race' to describe how the momentum inherent in population growth with its attendant problems for biodiversity is pushing us rapidly towards a catastrophic finish line (1986).

The rhetoric of 'catastrophe' was further enhanced through linking it to the fate of the dinosaurs.[6] In 1980, two eminent scientists from the University of California at Berkeley, Nobel prize-winning physicist Luis Alvarez and his geologist son Walter, proposed that the dinosaurs had perished as the result of climate changes brought about by an asteroid that had crashed on earth 65 million years ago. Few scientific theories have attracted more public interest as quickly as this controversial claim, a fact not lost on biodiversity activists who often used the dinosaurs as a point of comparison (Mazur & Lee 1993: 703). Similarly, a television advertisement sponsored by the Humane Society of Canada in the mid-1990s, proclaimed: 'It is the greatest extinction rate since the end of the dinosaurs.'

A subsidiary idiom here is that of 'entitlement'; thus, both Raven, in his testimony before the 1981 Congressional committee, and the IUCN, in the introduction to *World Conservation Strategy*, reiterate a memorable slogan to the effect that 'we have not inherited the Earth from our parents, we have borrowed it from our children' (see Box 8.1).

---

*Box 8.1 A sample of rhetorical statements on biodiversity
loss by prominent environmentalists / scientists*

'We have not inherited the Earth from our parents, we have borrowed it
from our children.'

> Peter Raven in Congressional testimony (see Kellert
> 1986) and IUCN World Conservation Strategy,
> Introduction (1980)

'We are in a race. Maybe we should call it a contest.'

> Brown (1986)

'We're treating the world as a business in the process of liquidation.'

> Peter Raven, cited in Gooderham (1994)

'The future well being of human civilization and that of many of the 10 million
other species that share this planet hangs in the balance.'

> McNeely *et al.* (1990b)

'In the last twenty-five years or so, the disparity between the rate of loss
and the rate of replacement [of species and populations] has become alarming;
in the next twenty-five years, unless something is done, it promises to become
catastrophic for humanity.'

> Ehrlich and Ehrlich (1981)

'Elimination of lots of lousy little species regularly causes big consequences
for humans, just as does randomly knocking out many of the lousy little rivets
holding together an airplane'

> Diamond (2005)

Running parallel to this 'doomsday' rhetoric (see Chapter 1) is a second type of
claims language that stresses the positive economic benefits of preserving diverse
habitats. Using warrants that are loaded with financial figures, proponents favour
a 'rhetoric of rationality' (Best 1987). For example, just over a decade ago, Walter
Reid, a vice president of the World Resources Institute, and colleagues, wrote this
about the 'economic realities of biodiversity':

> Currently some 25 per cent of US prescriptions are filled with drugs
> whose active ingredients are extracted or derived from plants. Sales of these

plant-based drugs amounted to $4.5 billion in 1980 and an estimated $15.5 billion in 1990. . . . In Europe, Japan, Australia, Canada and the United States, the market value for prescriptions and over-the-counter drugs based on plants was estimated to be $43 billion in 1985.

(1993–4: 49)

Significantly, this rhetoric uses the language of frontier development, for example, referring to 'bioprospecting' (Eisner 1989–90; Reid *et al.* 1993) or 'biotic exploration' (Eisner & Beiring 1994). It is suggested that somewhere in the 'biotic wilderness' scientists will find an equivalent of Madagascar's rosy periwinkle with its famous cancer-fighting properties (Eldredge 1992–3: 92).

This depiction of tropical rainforests as the cradle of tomorrow's pharmaceutical medicine spilled over into the arena of popular culture. In the American motion picture *Medicine Man*, Sean plays a maverick who discovers a miracle cure for cancer among the canopies of the South American rainforest, only to have his research site flattened by the bulldozers of a road-building crew. In *Day of Reckoning*, a 1994 action movie with a 'new age' flavour, an adventurer hunts for a rare plant with medicinal powers in the rainforests of Burma. As W.H. Hudson's romantic novel *Green Mansions* illustrated nearly a century ago, the human threat to the diversity of life in tropical ecosystems can make a compelling drama, with strong moral and spiritual overtones.

Contesting the claim

While individual countries undertook unilaterally to protect endangered species and habitats, it became obvious at least 35 years ago that concerted global action on biodiversity loss required some type of coordinated multilateral agreement. In fact, an International Convention on Biological Diversity was first proposed in 1974 and active planning for such an accord began in earnest in 1983 (Tolba & El-Kholy 1992). This process culminated in 1992 with the preparation of a Global Biodiversity Strategy under the auspices of three agencies: the United Nations Environment Programme (UNEP), the World Conservation Union (IUCN) and the World Resources Institute (WRI). In order to carry out the recommendations of this strategy, a Convention on Biological Diversity was introduced at the United Nations Conference on Environment and Development (UNCED) in Rio in June, 1992. By all accounts, this convention was conceived in a medium of considerable controversy, especially with regard to the question of access to genetic resources in Southern hemisphere nations.

At the core of the treaty can be found a basic tension between two conflicting commitments. On the one hand, the proposers wished to provide a mechanism whereby the international conservationist community could directly intervene in situations where sensitive environmental areas with diverse biological resources were threatened, notably in the tropical rainforests of Brazil. On the other hand, target nations weren't eager to lose their national autonomy and give up the right to make their own decisions, particularly with regard to development projects. As compensation for allowing outsiders to infringe their traditional national sovereignty, less developed nations wanted something in return, specifically, financial resources and the transfer of technology from the industrial nations of the North.

Furthermore, the Southern nations wanted to use the occasion to tighten up access to their genetic resources that theretofore had been more or less free to all comers. According to Article 15 of the Convention, nations were to have sovereign rights over their genetic resources and grant access only on mutually agreed terms and with 'prior informed consent'. Other provisions attempted to deal with some of the more potentially exploitative aspects relating to the appropriation of Third World genetic resources by multinational biotechnology companies. Increasingly, these firms have begun to prospect tropical habitats for unusual species of plants and animals, to 'borrow' key genetic material, bioengineer and patent it and then license the improved product back to the country of origin at a hefty profit. Accordingly, the South argued for access to the results and benefits of biotechnologies developed in connection with those genetic materials that have been exported, specifically in the form of continuing royalties and technology sharing.

Even at the pre-summit stage, a number of these points were contested. For example, an earlier draft of the convention had called for two global lists – a Global List for Biological Diversity and a Global List of Species Threatened with Extinction on [a] Global Level – that would have spelled out in priority fashion the commitments that were required of signatories to the treaty. However, during the final negotiations at Nairobi, Kenya, leading up to the Rio Summit, these references to global lists were removed, a measure that was strongly contested by many delegates including the leader of the French delegation who refused to sign the final act. Similarly, a provision that would have furnished free 'scientific access' to genetic resources in biologically diverse nations was dropped from the final convention (Barton 1992).

At the summit itself, the United States incurred the wrath of other participants by refusing to sign the Biodiversity Convention, even though 153 other countries did so and the Secretary of the Environment himself was in favour. This appeared to be the result of considerable pressure on President Bush from American biotechnology trade associations, which objected to the provisions that would have meant that US firms must pay continuing royalties and share new patents and technological secrets with nations whose biological resources are the source of new products (Susskind 1994: 182).

The Biodiversity Convention was challenged by a third party who wasn't present at either the negotiations or the Summit. This was a coalition, of farmers, ecological activists and others from Southern nations who felt that local people had been excluded from the formulation of the treaty, especially the provisions relating to intellectual property rights. Their absence has subsequently been noted annually in the *Report of the Global Biodiversity Forum* 'Background' section, with the statement: 'However, the process prior to and following the development of the CBD [Convention on Biological Diversity] did not in general allow for the full participation of all those interested and affected.'

The best-known spokesperson for this movement is the Indian eco-activist Vandana Shiva. Shiva and her movement have attempted to wrest 'ownership' of the problem of biodiversity loss from conservation biologists, non-governmental global environmental organizations and government negotiators who they accuse of assuming a mantle of leadership that is not theirs to wear. In particular, they object to the exclusion of the original donors of genetic resources – Third World farmers – from the exchange of resources and knowledge that the Convention governs. The basic problem, Shiva states, is that

> those 'selling' prospecting rights never had the rights to biodiversity in the first place and those whose rights are being sold and alienated through the transaction have not been consulted or given a chance to participate.
>
> (1993a: 559)

Shiva observes that even in the case of the 1991 agreement between Merck Pharmaceuticals and INBIQ, the National Biodiversity Institute of Costa Rica, a much-heralded and publicized example of how it is possible for multinational corporations to compensate the Third World for its genetic resources, the people living in or near the national parks in Costa Rica were not consulted, nor were they guaranteed any economic benefits. Rather, the agreement was forged between Merck and a conservation group formed at the initiative of a leading American conservation biologist Dan Jantzen, who, it will be recalled, is a member of Wilson's 'rainforest mafia' (Shiva 1993: 559).

Opponents of 'commercialized conservation' (Shiva 1990: 44) have proposed the formulation of an alternative form of intellectual property, the *Samuhik Cyan Sanad* or Collective Intellectual Property Rights (CIPRs). These collective patents invest the right to benefit commercially from traditional knowledge in the community that developed it. Furthermore, it is demanded that multinational companies seeking to utilize Third World genetic resources be compelled to deal through the village organizations who would hold title to these CIPRs. Failure to do so, it is claimed, would constitute 'intellectual piracy' (Shiva & Holla-Bhar 1993: 227).

Shiva's challenge did not go unnoticed. At the 7th Session of the Global Diversity Forum, held in Harare, Zimbabwe in June, 1997, the official 'Statement' prepared by Forum participants included the following paragraph:

Participants recommended that CITES mechanisms be developed to incorporate local and traditional knowledge and local participation in decision-making at all levels including in the national scientific bodies and international forums. National governments should be encouraged to involve local communities in the development and implementation of CBD and CITES strategies. All parties should be encouraged to include assessments of potential impacts on local communities when proposing changes to existing Conventions. Improvements are needed in the national and international processes for carrying out the goals of the Conventions to reflect the rights and aspirations of local communities.

(Global Biodiversity Forum 1997)

Most recently, international biodiversity discourse has shifted towards the reconciliation of conservation and poverty reduction through development, two goals that have often been depicted in the past as being at odds with one another and driven by different moral agendas. Thus, the IUCN's director general now describes protected areas as 'islands of biodiversity in an ocean of sustainable development' (cited in Adams *et al.* 2004: 1146). In addition to this, the Millennium Development Goals (MDGs), adopted at the 2000 UN Millennium Summit, links environmental sustainability with poverty eradication, education, gender equality, reduced child mortality, environmental sustainability and the creation of a global partnership for development. Conservation and the sustainable use of biodiversity are becoming increasingly perceived as critical to the full achievement of the MDG goals (Timmer & Juma 2005: 27).

One high-profile programme that has attempted to convert this rhetoric into achievable gains is the *Equator Initiative*, launched in 2002. The initiative is a partnership among local grassroots groups in countries along the equatorial belt, the United Nations and the UN's global development network. Its centrepiece is the Equator Prize, awarded to local community partnerships that work simultaneously toward sustainable income generation and environmental conservation. Some past prize recipients include the Green Life Association of Amazonia (AVIVE) in Brazil, which focuses on the sustainable extraction and marketing of medicinal and aromatic plant species; the Genetic Resource, Energy, Ecology and Nutrition (GREEN) Foundation in India, which works through a network of women's farming groups called '*sanghas*' to improve food security by conserving indigenous seeds and establishing community seed banks and home gardens; and the Suledo Forest Community in Tanzania, which harnesses local knowledge of the forests to regulate poaching and promote sustainable silviculture (Timmer & Juma 2005).

Despite the promise of such prizewinning projects, the dual goals of biodiversity loss and poverty eradication are not always compatible. Projects that seek to integrate conservation and development have tended to be 'overambitious and under-achieving' and lasting positive outcomes remain 'elusive' (Adams *et al.* 2004: 1147).

For example, ecotourism ventures, one popular type of project undertaken by Equator Prize winners, are risky insomuch as they are vulnerable to international tourist fads; create pressure to build hotels and leisure facilities that negatively impact the resource base on which the community depends; and may fail when other local communities choose ecotourism as their source of alternative livelihood, thereby saturating the market (Timmer & Juma 2005:31–2). Other projects falter when they are hijacked by local elites as a way of solidifying their interests. As Timmer and Juma (2005: 35) note,

> ignoring differences in values, perspective and power within a community and [the] differential access that community members have to layers of political decision makers leads to inaccurate assumptions about the ease by which collective decisions at the local level can be made.

Most recently, a schism has torn the fabric of biodiversity conservation. In January, 2009, Friends of the Earth International (FOEI) withdrew their membership of the International Union for the Conservation of Nature, citing concern with the corporate partnership between Shell, the multinational energy company, and IUCN. MacDonald (2010: 514) reports that this is 'but one example of a growing ideological and material divide between large international conservation organizations and smaller groups that orient themselves around the grassroots'.

In MacDonald's account, major changes surfaced around 2000, when conservation organizations began to seriously engage with the private sector. To a certain extent this shift is tactical, insofar as the leadership believes that a partnership with business can open up opportunities to influence corporate behaviour. Some political ecologists have coined the term 'Nature™ Inc.' to shorthand the intrusion of market forces into environmental policy and biodiversity conservation (Arsel & Büscher 2012).

Furthermore, the development of 'business and biodiversity' initiatives has been influenced by political and fiscal considerations. During the 1980s, national governments, in many cases under pressure from multilateral financial institutions such as the World Bank and the International Monetary Fund, reduced their investment in long-term biodiversity protection, especially in the Global South. This created a promising new political opportunity for large conservation organizations. Through the 1980s and 1990s these agencies rapidly expanded in size and budget and 'engaged in a "mission shift" moving from a focus on knowledge production and policy consultation to fund raising and project implementation' (MacDonald 2010: 520). To finance this expansion, conservation organizations increasingly turned to corporate donations[7] and to 'market-based incentive projects' such as trophy hunting, bioprospecting and ecotourism. Increasingly, this new phenomenon of globalized neoliberal conservation is coming into conflict with ways of understanding natural resource management and biodiversity that revolve around local needs and identity (see Box 8.2).

Box 8.2 Peace parks in Southern Africa

One increasingly popular way of protecting African biodiversity while promoting regional cooperation and development is the International Peace Park or Transfrontier Conservation Area (TFCA). Established and managed by two or more countries, there are currently over 200 of these worldwide, many of which are located in Southern Africa (Büscher 2010: 35). One of the most complex of these is the Maloti-Drakensberg Transfrontier Conservation and Development Project (MDTP), a World Bank/Global Environment Facility funded initiative situated between South Africa and the tiny neighbouring nation of Lesotho.

The Dutch social scientist Bram Büscher conducted ethnographic research on MDTP between 2005 and 2007. He found a sharp conceptual disparity between the partners in the intervention. The Project Coordination Unit (PCU) (implementing agency) in South Africa favoured Bioregional Conservation Planning (BCP). Deriving mostly from the natural sciences, this approach emphasizes technical expertise in the protection of biological diversity. The value for people is indirect and long term, deriving from a more constructive balance between human needs and the conservation of biodiversity (Büscher 2010: 38). The Lesotho PCU approach is more local in scope, emphasizing community development among small commercial farmers and village residents. Deriving chiefly from the social sciences, it puts primacy on people. Resource conservation is ideally about the economic or use value it brings to local people (Büscher 2010: 38). Broadly, the South African approach hinged more towards 'globalism', while the Lesotho approach could be termed 'Africanist' (Büscher & Whande 2007).

On the ground, the two PCUs clashed sharply and often, competing for legitimacy and acceptance within the wider conservation and development marketplace. The Lesotho PCU often felt like a junior partner. The South African PCU challenged their data as 'anecdotal' and placed minimal importance on the efforts of the Lesotho PCU to get as many local stakeholders as they could to buy into the project. A year into the MDTP, interpersonal relations between staff members on both sides of the project had become so strained that two external mediators were brought in. When an external evaluation ranked the South African PCU higher than its Lesotho partner, a diplomatic crisis ensued.

Büscher and his colleagues conclude that international peace parks, promoted as a 'win–win' solution that promotes international cooperation and regional development 'happens not only in spite of troubling contradictions and problems, but indeed because of them' (Büscher 2013a). In fact, they have become part of 'the politics of neoliberal conservation'.

MDTP, for example, was conceived in part as an incentive for overseas visitors to the FIFA 2010 World Cup to travel further afield than the urban areas of Cape Town, Durban, Johannesburg and Pretoria. Conservation and development interventions can therefore be regarded as 'highly politicized constructions' (Büscher 2010: 48).

Conclusion

The rapid ascent of biodiversity loss in the international arena is rather surprising. While extensive, research on biodiversity largely navigates uncharted waters. Of the 1.4 million species known around the time that the Convention on Biological Diversity was adopted (this has since risen to approximately 1.75 million [Spray & McGlothlin 2003a: xvi]), only five per cent can be considered 'well known' and the relationships between many of them are a mystery (Gooderham 1994: A-12). The theory that underlies ecosystem diversity is based primarily on small-scale studies of ponds projected on the larger screen of nature. The benefits of acting boldly are not precisely documented. The costs are considerable, not only financially, but also in behavioural terms. If large-scale biodiversity protection is to be implemented, the number and range of people affected and the extent of change required are considerable (Balch & Press 2003: 124–5). The ownership of the problem is disputed with multiple claimants.

Despite these drawbacks, biodiversity became a major environmental theme in the 1990s. There are several factors that account for this.

First, it is not purely an environmental problem but is simultaneously an economic and political question. For national states, especially in the South, it can be both a source of foreign exchange and a window through which First World biotechnology can be accessed. For entrepreneurs biodiversity can be transformed into a valuable resource that can generate a tidy profit. Even more significantly, for the corporate sector, business and bio-diversity initiatives contribute to a broader strategy of preventing framework conventions such as the Convention on Biological Diversity from imposing tough regulatory limits on their access to 'nature as capital' (MacDonald 2010: 525).

Second, biodiversity loss constitutes a socially constructed environmental problem that has brought together two well-established organizational sectors: the international development establishment and the global conservation network. Nested in a web of NGOs, notably those that connected to the United Nations, it has an institutional momentum extending beyond that which is able to be generated by single environmental movement organizations such as Greenpeace and Friends of the Earth, which have more of an 'outsider' status.

Third, the biodiversity problem has not been constructed from scratch but has flowed out of the already long-standing problem of endangered species. The two problems are to a large extent symbiotic and synergistic. Biological diversity gives species endangerment and extinction a theoretical grounding that it previously lacked. The example of endangered species provides biological diversity with a specific focus and an emotional resonance that the more general issue often lacks. Furthermore, the preservation of diversity furnishes a rationale for action in rancorous environmental disputes such as those that have raged in recent decades over Great Whale River and the Clayoquot Forest in the Canadian North and West (Suzuki 1994b)

Finally, the location of biological diversity at the centre of the discipline of conservation biology means that it has been buffered against the 'issue attention cycle' (Downs 1972) that affects a great many other environmental issues. Furthermore, 'the biodiversity debate has not been embroiled in the kind of scientific disputes that have occurred in debates on acid rain, the depletion of the ozone layer and climate change' (Valiverronen 1999: 407). Conservation biology provides biodiversity loss with a centre of gravity around which it can revolve, rotating out into the realm of international diplomacy and conflict but stabilized by the continual pull of research within this specialty area.

Notes

1   At the first official SCB meeting in April, 1988, many participants cited the need for aggressive conservation action rather than research as the top priority (Tangley 1988: 444).

2   By contrast, *taxonomy*, a specialty science that involves identifying and cataloguing biological species, has been in steady decline for decades. Perceived to be a nineteenth-century descriptive science with little present-day application (Burton 2003), taxonomy was unable to claim ownership of biodiversity as an environmental problem, despite its vital importance in compiling species lists.

3   My chronology of these international conventions draws primarily on 'Annex 3: international legislation supporting conservation of biological diversity' in McNeely *et al.* (1990a).

4   Wilson's efforts to champion sociobiology have brought him acclaim, but have also generated considerable controversy, especially from some on the American political Left. In a now-famous incident at the 1978 meeting of the American Association for the Advancement of Science, demonstrators chanted 'Racist Wilson you can't hide, we charge you with genocide!' followed by someone pouring a jug of ice-water over his head shouting 'Wilson, you are all wet' (Shermer 2000).

5   The Northern spotted owl became one of the 'most celebrated and vilified endangered species' (Grumbine 1992: 144) in recent memory. With a habitat and geographic range that stretches the length of old growth forests from British Columbia to northern California, protecting it under the Endangered Species Act implied a significant reduction in logging activities in the ancient forests. In the course of a decade of political and legal wrangling, the Northern spotted owl became a symbol for some of the unrealistic features of the Act.

6 This appears to have been a two-way street. Not only did the fate of the dinosaurs provide a powerful magnet by which diversity activists could attract the attention of the public, research on the immediate threat of extinction has also proven useful in understanding what happened 245 million years ago. For example, Niles Eldredge, in writing his book *The Miner's Canary: Unraveling the Mysteries of Extinction* (1991), relied heavily on Edward Wilson's published data and arguments to examine the relationship between the mass extinctions of the geological past and the present-day biodiversity crisis (Eldredge 1992–3: 90).

7 In 2006, for example, 56.5 per cent of Conservation International's 2006 budget was drawn from corporations and foundations. Some of the latter (the Walton Family Foundation) derive their funds from corporate earnings (WalMart).

Further reading

Büscher, B. (2013) *Transforming the Frontier: Peace Parks and the Politics of Neoliberal Conservation in Southern Africa*, Durham, NC: Duke University Press.

Shiva, V. (1993b) *Monocultures of the Mind: Perspectives on Biodiversity and Biotechnology*, London: Zed Books.

Wilson, E.O. (ed.) (1986) *Biodiversity*, Washington, DC: National Academy Press.

Online resource

USAID, Conserving Biodiversity and Forests. Available HTTP: http://www.usaid.gov/what-we-do/environment-and-global-climate-change/conserving-biodiversity-and-forests

9　Fear of fracking

In May, 2013, the Association of German Breweries voiced its grave concern to the federal cabinet that drilling for shale gas using the method of hydraulic fracturing or fracking could violate the 500-year-old Bavarian purity law by endangering the water that more than half Germany's breweries currently take from private wells (Nicola 2013). Drawn up in 1516 at the initiative of Duke Wilhelm IV, the *Reinheitsgebot* (Bavarian Purity Law) is the longest standing non-religious food safety regulation in the world today. A minor element of a more extensive document designed to control prices, the Law stipulated that only three ingredients, barley, hops and water may be used in brewing beer.[1]

Fracking is widely touted as the key to an energy revolution that would render conventional methods of extracting fossil fuels much less strategically important. Soaring domestic production of shale gas (and oil) have sent American energy imports plummeting, finally allowing the nation to pay some attention to fighting climate change (Reguly 2013). *The Economist* estimates that shale gas could contribute nearly half of America's gas supplies by 2035, up from a third at present. If this could be replicated worldwide, fracking would play an important role in providing energy to such nations as Australia,[2] China and South Africa (*Economist* 2012). Shale-gas basins have also been identified in France, Germany, the Netherlands, Poland, Romania, Russia, Spain and the Ukraine. According to the US Energy Department, at the beginning of 2012 estimated shale gas reserves in the United States totalled 862 trillion cubic feet, 34 per cent of total gas supplies. In China, shale formations hold even more, an estimated 1,275 trillion cubic feet of gas, 12 times larger than that nation's conventional fields (Carroll & Polson 2012). More than half of non-US shale gas resources are concentrated in five countries – China, Argentina, Algeria, Canada and Mexico ('Shale oil and shale gas resources are globally abundant' 2013).

Shale gas (and oil) drilling is anything but benign. Among the dangers attributed to fracking is the toxic contamination of drinking water, chronic illness, the poisoning of agricultural land, disastrous explosions and even the triggering of minor earthquakes. At a 2012 medical conference on hydraulic fracturing, delegates recommended that the United States should declare a moratorium on fracking until the health effects are better understood ('US doctors call for moratorium' 2012).

Until recently, conflict over fracking has played out almost exclusively in town halls, high school auditoriums and courthouses across rural America. A *New York Times* profile of Cooperstown, a town in upstate New York best known as the home of baseball's Hall of Fame, depicts this local debate over horizontal drilling as 'close and nasty', pitting neighbour against neighbour.

Times have been tough in many of these regions, especially in the wake of the mortgage crisis that has devastated the US economy over the past decade. Applebome (2011) cites the relentless growth of 'pastoral poverty' as a factor contributing to the willingness of people to surrender drilling rights on their land in return for cash payments and royalties. Property owners are initially visited in their homes by 'landmen', representatives of oil and gas companies who attempt to convince them to sign leases. Many have done so, either individually or as part of a landowner's consortium coalition.

As is the case with prior community conflicts, notably over the construction of Walmart and other big box stores, opponents were initially drawn disproportionately from the ranks of retirees and second-home owners from the big city. Later on, once serious environmental health problems begin occurring, the opposition spreads to 'accidental activists' (Wilber 2012: chapter 5), homeowners who signed leases with the energy firm but had come to experience bitter disillusionment.

Although thousands converged on the statehouse in Columbus, Ohio in June, 2012, there have been no national scale marches and mass demonstrations in Washington similar to those witnessed in relation to the Keystone XL pipeline. Nevertheless, once word of what was going on filtered out, national environmental groups with full-time staff and significant budgets, notably the Natural Resources Defense Council, Earthjustice, Clean Water Action and Earthworks, took up the fight (Wilber 2012: 181).

In the United Kingdom, fracking is in its infancy compared to the United States. The only company that has done any experimental drilling is Cuadrilla Resources, whose shares are partially owned by Centrica, parent company to British Gas ('Fracking: energy futures' 2013). However, with a recent report from the British Geological Survey suggesting unexpectedly large deposits of shale gas, the issue has suddenly been put into play politically. This has escalated further by the 'Battle for Balcombe', an anti-fracking protest in a wealthy village a train ride south of London that attracted a clutch of celebrities (Marina Pepper, Bianca Jagger) and turned into an eco-cause célèbre in the global media (Waldie 2013).

Unlike in the United States, where support for shale gas extraction tends to follow party lines, the divide in the UK is more complex. The ruling Conservative/Liberal Democrat coalition is divided between a pro-growth wing and an environmental wing. The former enthusiastically embraces fracking, seeing it as the key to an energy security and industrial expansion. Its influence is evidenced in Prime Minister David Cameron's widely quoted comment, 'Britain must be at the heart of the shale gas revolution', and his decision to allow generous tax breaks for shale gas producers. The latter faction has serious qualms, because it fears that fracking would industrialize the countryside and make it difficult to justify subsidies for renewable energy projects. To further complicate the political kaleidoscope, there are Tory Members of Parliament from rural constituencies who tend to be hostile to renewable energy projects such as wind farms, and are scarcely more inclined to 'warm up to the development of a shale gas field in their backyards' ('Fracking: energy futures' 2013). This is especially the case for the 'stockbroker belt', the rolling countryside in Surrey and Sussex south of London where an estimated 700 million barrels of recoverable shale oil have been forecast. The advent of drilling near the green lawns of mansions in the Wessex and Weald basins 'may widen the nation's shale-energy debate, which has focused on gas in northwest England' (Bakhsh 2013).

In Canada, shale gas development is burgeoning in the Western provinces: Alberta, British Columbia, Saskatchewan. In the East, it is considered more controversial and divisive: 'Shale-gas-rich Quebec has slapped a moratorium on fracking, while Nova Scotia and New Brunswick are hamstrung by public backlashes, which has made exploiting relatively low reserves politically unappealing' (Hussain 2012). There is considerable legal variation from province to province over ownership of drilling rights. In Alberta, homeowners do not have any legal claim to sub-surface mineral rights, while in Ontario they do. As a nation whose economy rises and falls in concert with the fossil fuels industry, national political support for fracking might be expected. However, this is not entirely the case, insofar as increasing production of shale energy means reduced demand for Canadian gas and oil, from both conventional and unconventional (notably the Alberta tar sands) sources.

In Australia and New Zealand, the issue of fracking tends to follow the contours of established environmental politics. In Western Australia, the Liberal Party actively encourages horizontal drilling, the Labor Party is committed to environmental assessment and regulation and the Greens have called for a moratorium. Anti-fracking groups such as 'No Fracking Way' and the Lock the Gate Alliance have forged a coalition with climate change activists such as 350 Australia, the Australian Youth Climate Coalition (AYCC) and the Global Climate Initiative. This has led to a much greater emphasis on shale/coal fracking as contributors to global warming than is the case in North America.

South Africa is thought to have the fifth-largest natural gas reserves in the world, notably in the Karoo, a semi-desert, ecologically sensitive region of the Eastern Cape. Unlike in the United States, underground rights belong to the government rather than individual residents. This means 'farmers have little financial incentive to welcome fracking or turn against their neighbors who oppose it' (Flanagan 2012).

In early 2012, the national government imposed a moratorium on hydraulic fracturing, but lifted it 18 months later. Opposition to fracking in the region has been led by the grassroots Karoo Action Group (TKAG). Its founder, Jonathan Deal, won a 2013 Goldman Prize, sometimes known as the Nobel Prize for the environment (Box 9.1).

A short history of fracking

If anyone could be said to be synonymous with fracking, it is Halliburton Company, a global oil well services corporation with operations in more than 80 countries. In 2011, Halliburton posted revenues of US$25 billion, most of which came from domestic shale explorations in the United States (Sourcewatch). Halliburton didn't invent hydraulic fracturing. That distinction is generally accorded to Floyd Faris, an employee of Stanolind Oil and Gas Corporation, part of Standard Oil of Indiana, who was granted a patent for hydraulic fracturing in 1946. Three years later, Haliburton secured an exclusive license to Faris' process and conducted the first commercial fracturing treatments in Oklahoma and Texas.

The basic method of hydraulic fracturing is relatively straightforward. Highly pressurized, chemically treated water and sand are injected deep underground, thousands of metres down a vertical shaft at high pressure in order to crack open tightly packed rock formations, thereby releasing gas deposits previously locked into the shale beds. This is preferable to drilling a conventional well, insofar as the latter would only tap into a small area right around the well. However, drilling vertically yields only limited amounts of gas. Shale formations typically extend hundreds of kilometres across but are less than 100 metres thick, 'hardly worth sending a vertical well into' (Ehrenberg 2012).

In the 1970s, energy companies partnered with the US government to develop new techniques to reach natural oil and gas deposits in underground rock formations (Giles 2012). Progress was painfully slow. In a catastrophic event in 1982 labelled 'Black Sunday', Exxon cancelled a massive shale oil[3] project in northwestern Colorado, and, soon after, the Reagan administration deemed shale oil production unfeasible and cancelled its federal synthetic oil programme (Tolmé 2010). At the same time, the easy-to-capture, shallow reserves of natural gas were exhausted, leading many in the industry to declare that the resource in the confines of the continental United States was dead (Zager 2010).

Two developments in the 1990s changed all of this.

First, George Mitchell, a wealthy American petroleum engineer and oil baron, discovered a way to extract methane gas from shale at a manageable cost. It took Mitchell and his engineers a decade to figure out how to exploit the Barnett shale formation in north Texas near Fort Worth, transforming it into one of the most prolific natural gas fields in the country (Bogan 2009). Mitchell's technique is called directional or horizontal fracturing. As the name suggests, it requires that

Box 9.1 Fracking warriors

Although he had no prior training in grassroots organizing, Jonathan Deal has become the leading opponent of plans by Royal Dutch Shell to drill for shale gas in the Karoo, a desert-like rural area in South Africa renowned for its natural beauty. Deal is a writer, photographer and nature lover who wrote a book, *Timeless Karoo*, in 2007. Upon reading an article in the local newspaper about Shell's plans to apply for exploratory drilling permits, he started a Facebook group to educate the public about the risks of fracking. This morphed into 'Treasure the Karoo Action Group' (TKAG). As TKAG's chair he led a team of scientists, legal experts and volunteers in preparing a comprehensive 100-page report to President Jacob Zuma that called for a moratorium on fracking. The group also lodged a complaint with the Advertising Standards Authority of South Africa about Shell's public relations campaign; the regulator found Shell guilty of four counts of false advertising for its claim that hydraulic fracturing had never caused any water contamination. In 2013, Jonathan Goldman was awarded the Goldman Environmental Prize (Africa) honouring grassroots environmentalism. That same year, he was chosen as the subject of a documentary to be shown on the PBS series *The New Environmentalists*, which is described as profiling 'ordinary people who effect extraordinary change'.

Another dedicated 'fracking warrior' is Sandra Steingraber, a biologist, poet and cancer survivor who has written a trio of books in a personal style about industrial pollution, health and the environment: *Living Downstream: An Ecologist Looks at Cancer and the Environment* (1997); *Having Faith: an Ecologist's Journey to Motherhood* (2011) and *Raising Elijah: Protecting Our Children in an Age of Environmental Crisis* (2011). In recent years, Steingraber has become a front line campaigner against fracking in the Finger Lakes district of upstate New York. While this has put her on Bill Moyers' public affairs television programme, it also landed her jail in Elmira, NY for blocking the driveway of Inergy Midstream, an energy and storage transportation company that plans to pump highly pressurized gas from the fracking fields into old salt caverns along and under Seneca Lake, which provide drinking water for 100,000 area residents.

Sources: Butigan (2013); Flanagan (2012); *BBC News Africa* (2013); 'Jonathan Deal warns against fracking' (2013)

wildcatters change the direction of the drill bit travelling downward in the well, executing a 90-degree turn (this is called 'kicking off') and moving it horizontally through the shale formation. Millions of gallons of water and fracking fluid, together with fine sand, are then pumped into the well at high pressure, fracturing the shale all along its area. The fractures in the shale are kept open long enough by the sand to allow the gas to flow out and up to the surface. Mitchell's fracking technique rejuvenated a dying industry and led to 'something as sizeable as a gold rush' (Zager 2010).

There is just one catch – horizontal fracking may be polluting our drinking water. This can happen in two ways. Deep underground, the blasts that liberate gas from the shale formations may also create unexpected pathways for gas or liquid to travel between deep shale and shallow groundwater (Mooney 2011). Above ground, the chemically laden 'flowback water' that comes back up after wells have been fractured is usually stored in open-air pits. When these pits are improperly lined or overflow in heavy rains, the toxic soup from hydraulic fracturing escapes into the watershed. Indeed, major river basins such as the Susquehanna and Delaware appear to have been contaminated because of inadequate handling of flow-back water (Mooney 2011).

With this potential environmental hazard hovering, energy companies needed a legal shield. Enter once again the Halliburton Company in the person of Dick Cheney, who had served as Halliburton's CEO after he was defense secretary under George Bush. At Cheney's request – he had subsequently become vice president of the United States – a two-paragraph clause known as the 'Halliburton Loophole' was inserted into the Safe Drinking Water Act. Among other things, this did away with lengthy environmental review on Bureau of Land Management properties and shielded energy companies from having to reveal the chemical content of fracking fluids.

Constructing fracking as an environmental problem

Assembling the claim

As I noted in Chapter 3, in the initial 'assembly' stage of issue construction, a central task is to name the problem, diagnose its origins and symptoms and distinguish it from other similar or more encompassing problems. At first glance, fracking appears to be merely the latest chapter in the 'enduring conflict' (Schnaiberg & Gould 1994) between the treadmill of production and the natural environment.

However, there are several aspects of fracking that give it a distinctive discursive DNA as compared to other methods of oil and gas extraction. The word itself is onomatapoeic, meaning that it makes a noise that imitates the action or sound that it represents. Other examples of this are the words bash, crackle and smash. Carol Gray, shop assistant in Malton, a North Yorkshire agricultural community recently found to be located over a sizeable shale gas deposit, told a reporter for *The Independent*, 'I have heard of fracking. It sticks in your mind because it's such a horrible word' (Brown 2013). Commenting on a nationwide survey conducted by the Pew Research Center measuring support/opposition for fracking, Republican pollster David Hill (2013) suggests that most respondents probably could not explain what fracking actually is. Rather, they respond to verbal cues. For example, they 'revulse at the ugly "f-ing" sound of the term for hydraulic fracturing'.

One compelling visual image has come to define the perils of fracking. As he travelled around eastern Pennsylvania, documentary filmmaker Josh Fox kept hearing stories about flammable gas rushing out of taps in people's homes. One family's water well even exploded on Christmas Day. In December, 2007, a house in Bainbridge Township in northeast Ohio exploded when, investigators determined, natural gas from a well 1,000 feet away migrated into the home due to the pressurizing of the well's surface casing (Bachman 2011). Upon reaching the Western Slope in Colorado, Fox found a homeowner who was even willing to demonstrate this on camera – the fellow just barely escaped incineration when he lit a match, held it up to the tap in his kitchen sink, and abruptly leapt back to avoid the wall of fire. This flames-in the-faucet trope has recurred elsewhere. Jeff Goodell, who wrote a long article on fracking for *Rolling Stone* magazine, interviewed Sherry Varson, an Oklahoma dairy farmer who leased mineral rights beneath her farm to Chesapeake Energy Corp. in 2007. Goodell (2012) writes:

> 'I discovered I could light my water on fire,' she [Vargson] says. 'And I still can.' And to demonstrate, she walks over to the faucet in her kitchen, lights a match and turns on the faucet. Whoosh! A flame shoots out like a blowtorch.

On the website of ProPublica, which describes itself as pursuing 'journalism in the public interest', you can even watch a snappy music video entitled *The Fracking Song* (My Water's On Fire Tonight). This summons up memories of a notorious event in the 1960s when the Cuyahoga River in Cleveland, Ohio became so polluted with industrial waste that it actually caught fire.

Finally, there is the spectre of shale gas extraction terrorizing the countryside clad as an agent of seismic activity. Scientific support for this has varied. In 2012, a report by the US Geological Survey concluded that a remarkable increase in earthquakes in the United States since 2001 is 'almost certainly' the result of fracking ('Fracking for oil' 2012). According to John Filson, chief of earthquake studies at the Survey, dumping fracking fluid left over from the drilling process deep underground lubricates ancient faults, leading to a buildup in pressure ('Fracking did not cause East Coast quake, doubts linger' 2012). On the other hand, an analysis released by the US National Research Council in that same year concluded that fracking is not a key risk factor for quakes strong enough to be felt by people ('Little evidence to link fracking to earthquake risk' 2012).

Seismic activity linked to fracking has had a higher profile in some European nations. Concurrent studies commissioned by both the British Geological Survey and Cuadrilla Resources found that two small earthquakes that struck near Blackpool along the coast in northwest England were caused by the presence of water in rocks surrounding a shale gas deposit, although the two reports differ on the possibility of this happening again (Marshall 2011). Probably the strongest link between shale gas drilling and earthquakes is in the Dutch city of Groningen along the North Sea, where earthquakes have been averaging one a week (Waterfield 2013).

Currently, the public does not seem to view seismic activity as the most important risk related to fracking (water contamination and health issues dominate). In response to an open-ended question regarding the most important environmental risk associated with hydraulic fracturing, less than two per cent of respondents in a 2012 telephone survey conducted by the University of Michigan cited earthquakes (Brown *et al.* 2013: 12). Nevertheless, the idea of horizontal drilling generating earthquake activity is potentially powerful, insofar as it is vivid and visual, and evokes catastrophic images that are more difficult to imagine in relation to ground water contamination.

Unlike many other incipient environmental problems, claims about the dangers of fracking have not been constructed primarily within the realm of science. Amy Mall, a senior policy analyst with the Natural Resources Defense Council, describes fracking as 'a big experiment without any actual solid scientific parameters guiding the experiment' (Mooney 2011: 85). *Science News* staff writer Rachel Ehrenberg agrees with this assessment, noting, 'Despite all this activity, not much of the fracking debate has brought scientific evidence into the field' (Ehrenberg 2012). Writing in the *Scientific American*, Chris Mooney (2011), author of the bestselling book *The Republican War on Science*, observes, 'The scientists and regulators now trying to answer this complex question have arrived a little late. We could have used their research *before* fracking became a big controversy'.

Nadia Stenzer, an organizer for Earthworks, a Washington, DC-based environmental group, calls fracking 'a case of technology moving ahead of the science' (Giles 2012). This presupposes that those industry researchers who have contributed directly to the development of these advanced techniques of hydraulic fracturing do not generally see them as problematic, whereas less compromised scientists in university and environmental agencies are prone to take a more critical view. Writing in the leading scientific journal *Nature*, Robert Howarth and Anthony Ingraffea observe: 'Because shale-gas development is so new, scientific information on the environmental costs is scarce. Only this year have studies begun to appear in peer-reviewed journals, and these give reason for pause' (Howarth *et al.* 2011).

Nor did the national ENGOs (environmental non-governmental organizations) in the United States rush to position themselves in the vanguard of those opposing hydraulic fracturing. Some leading environmental groups initially welcomed fracking, in large part because they believed that it would sound the death knell for 'dirty coal'. Fred Krupp, head of the Environmental Defense Fund, evidently called the shale gas boom 'a potential game changer – a cleaner energy source that could replace coal and oil for a few decades, until the cost of wind and solar power dropped enough to put fossil fuels out of business' (Goodell 2012). In an exclusive investigative report, *TIME* magazine reported that the Sierra Club, the largest and oldest environmental organization in the United States, had accepted over US$25 million in donations between 2007 and 2010 from the gas industry, mostly from Aubrey McClendon, CEO of Chesapeake Energy, one of the biggest frackers in the world. In 2009, Carl Pope, executive director and CEO of the Sierra Club, accompanied McClendon on trips around the country aggressively promoting the benefits of natural gas over coal (Walsh 2012).

Recently, however, it has occurred to the environmental establishment in the United States that a fracking boom is more likely to discourage the development of alternative, green energy technologies, such as solar panels and wind turbines, than pave the way to their replacing carbon-based fuels. While generally responding positively to President Obama's June, 2013 Climate Action Plan, the Sierra Club criticized his positive stance toward fracking. 'When it comes to natural gas, the president is taking the wrong path', Deb Nordone, the head of the Club's 'Beyond Natural Gas' (re-branded from 'Beyond Coal') campaign wrote in a blog post (Begos 2013).

With scientists and the environmental movement effectively perched on the sidelines, claims about the dangers of hydraulic fracturing started in hundreds of local communities where drilling for methane has been happening for a decade with alarming consequences and minimal national awareness.

Presenting the claim

If there is a single identifiable source responsible for elevating fracking to a broad level of public awareness and concern, it is the Academy-Award-nominated documentary film *Gasland*, written and directed by Josh Fox. This is sometimes called the 'Gasland effect' ('Sorting frack from fiction' 2012: 8). *Variety*, the entertainment industry bible, calls *Gasland* 'one of the most effective and expressive environmental films of recent years'. Fox says the idea for *Gasland* began when he received a letter in the mail from a national energy company offering him nearly US$100,000 in return for leasing drilling rights on 19.5 acres of land that he and his family own in a wooded area along a tributary of the Delaware River. Unable to find out very much about the topic online, Fox set out to explore the proliferation of new wells along the Marcellus Shale in his home state of Pennsylvania. Encountering scores of people with horror stories about polluted drinking water, ill health and even an exploding well, he embarked on a cross-country odyssey, including a stop in Dish, Texas (Box 9.2).

Gasland hasn't earned much money in commercial movie theatres – its Worldwide Box Office is just under US$50,000. However, it has diffused widely through other media channels. I first learned about fracking whilst returning by air from the UK

Box 9.2 Dishing it out

Dish (formerly Clark), Texas is a town of 225 people, 35 miles north of Fort Worth. The people of Dish have been extraordinarily willing to sell their community to any and all corporate callers. In 2005, they even changed the name of the town as part of a deal to get free satellite TV service. More crucially, many residents agreed to permit energy companies to drill wells on their property using hydraulic fracturing. Within years, a host of health complaints ranging from nosebleeds to cancer were showing up. According to former mayor Calvin Tillman, half of those polled in a municipal health survey had a symptom such as itchy eyes, a bloody nose or a scratchy throat related to one of the chemicals cited in an air quality study commissioned by the town (Hamilton 2012b). Alas, efforts by residents to link these to the gas wells in their backyards have failed to establish a conclusive cause–effect relationship. In part, this is because the prevailing winds blow industrial and automotive pollution up from Dallas and Fort Worth, making it difficult to trace health problems to a single source. As for the mayor, he and his family left Dish for good after his young sons began to get serious nosebleeds in the middle of the night. Tillman now works for 'Shale Test', a nonprofit group that does testing for low income families and communities affected by natural gas drilling (Sturgis 2012).

---

to Toronto, where *Gasland* was on offer through Air Canada's in-flight entertainment system. The film won a special jury prize at the Sundance film festival and, subsequently, an Oscar nomination for best documentary film. Josh Fox has been interviewed on numerous newscasts and current affairs television shows, notably on ABC and on PBS. A sequel, *Gasland II*, was broadcast by HBO in 2013. Initial concern about fracking in South Africa was fuelled by a showing of *Gasland* in Cape Town by the environmental group Earthlife Africa (Flanagan 2012).

In May, 2012, National Public Radio in the United States broadcast a trilogy on the topic of fracking over a period of two days. The narrative presented here is one of methane gas leaks, industrial pollution and thousands of local residents falling ill. The main challenge faced by local communities, the producers claim, is to definitively prove that fracking is making people sick.

The first episode, 'Medical records could yield answers on fracking', focuses on a proposed study of residents of northern Pennsylvania that would look at detailed health histories of thousands of people who live near the Marcellus Shale, a rock formation in which energy companies have already drilled about 5,000 natural gas wells (Hamilton 2012a). As a first step, researchers propose starting with asthma patients. The Geisinger Health System, which provides care to more than two million

Pennsylvanians, maintains a huge database of electronic records, including information that would permit analysts to gauge how far each asthma patient lives from a well site. The second episode, 'Town's effort to link fracking and illness falls short', was broadcast on the NPR flagship public affairs programme *All Things Considered*. The third instalment, 'Fracking's methane trail: A detective story', features the research of Gaby Petron, an air pollution sleuth with the National Oceanic and Atmospheric Administration (NOAA). Using readings from an NOAA observation tower north of Denver, Colorado, Petron linked high levels of a chemical cocktail that included methane, propane and pentane to a nearby complex of gas fields where there were 20,000 active wells (Shogren 2012). Petron's study is notable because it indicates that far more methane gas is leaking from gas wells, storage tanks and equipment than industry representatives allow. An accompanying interactive site, 'Explore shale', provided background material on the natural gas drilling process and how it is regulated.

When producers of the 2012 Hollywood movie *Promised Land* were given the green light to film in the shale fields of Pennsylvania, they no doubt hoped it would rival *The China Syndrome* (1979) – negligence leading to a meltdown in a nuclear power plant – and *Erin Brockovich* (2000) – cover up of heavy metal groundwater contamination by a gas and electric company. With an edgy, acclaimed director, Gus Van Sant, and two popular actors, Matt Damon and John Krasinski, critical and box office success seemed assured. Alas, *Promised Land* failed to deliver. Undercut by a weak script, technical inaccuracies and an unsatisfying ending the film quickly disappeared from the multiplex.

Contesting the claim

On 25 June, 2013, US President Barrack Obama delivered a speech at Georgetown University in Washington, DC outlining his long-awaited Climate Action Plan. Promising to significantly reduce carbon emissions by reducing American use of oil and coal, he singled out natural gas as playing a lead role in the nation's energy renaissance. Echoing an argument frequently espoused by proponents of shale gas drilling, Obama characterized increased gas production as a bridge between the 'dirty' fuels of today and the low carbon, green economy of the future. Canadian journalist Eric Reguly explains:

> Since gas's carbon content is about half of that of coal, the American carbon footprint is shrinking, allowing Mr. Obama to take the moral high ground as carbon output rises in countries bent on polluting their way to prosperity, as North America and Europe did. Minus the shale revolution, his climate plans would have been somewhere between non-existent and window dressing. Shale gas is responsible for the carbon-reduction progress that the regulation-mad U.S. Environmental Protection Agency could never achieve on its own.
>
> (Reguly 2013: A-2)

Contesting the fracking issue has suddenly been made more difficult politically. Not only do opponents need to parry a 'patriotic' narrative that identifies shale gas extraction as the key to US energy sufficiency, they also need to counter an environmental argument that says increased shale gas consumption means lower carbon emissions and, therefore, reduced global warming.

As was previously the case with environmental justice issues, mainstream ENGOs have been playing catch-up, challenging the shale oil industry in the courts and council chambers of the nation. For example, a coalition of environmental advocacy groups (Grand Canyon Trust, Center for Biological Diversity, Sierra Club, Living Rivers, Southern Utah Wilderness Alliance, Biodiversity Conservation Alliance, Rocky Mountain Wild) in the Western states recently filed a 60-day intent to sue the Bureau of Land Management for failing to meet with the US Fish and Wildlife Service before passing new oil shale and tar sands regulations.

Like the protest against the proposed Keystone XL pipeline, shale oil drilling has attracted a groundswell of celebrity 'fractivists'. Especially influential here are Yoko Ono and Julian Lennon,[4] who founded a campaign, 'Artists Against Fracking', which includes among its members such high profile personalities as Alec Baldwin, Lady Gaga, Robert Redford, Salman Rushdie and Susan Sarandon. In a radio ad, Redford argued, 'Fracking is a bad deal for local communities. It's been linked to drinking water contamination all across the country', while Baldwin, in an editorial in the *Huffington Post*, described a scenario in which energy companies promise people 'some economic benefit, deliver a pittance in actual compensation, desecrate their environment and then split and leave them the bill' (Begos & Peltz 2013).

In a seminal article written nearly 15 years ago, William Freudenburg (2000) called for increased attention to the 'social construction of *non*-problematicity' of environmental problems. By this, he meant the process whereby powerful interests in industry and government re-define and de-legitimate conditions defined by environmental claims-makers in science, media and the community as worrisome. In so doing, they 'succeed not only in keeping environmental problems and technological risks off the political agenda, but also in preventing such conditions from becoming widely defined as problems in the first place' (McCright & Dunlap 2003: 352).

Not surprisingly, energy companies contend that hydraulic fracturing methods including horizontal drilling pose a minimal risk. Brian Grove, Chesapeake Energy's director of corporate development, told *Rolling Stone* that the layer of shale being drilled into the Marcellus Shale is 7,000 feet beneath the surface, whereas drinking water rarely runs deeper than 1,000 feet. 'That leaves 6,000 feet of rock in between', he asserted: 'There is no way that any fluids are going to migrate up from the shale rock to the drinking-water aquifers' (Goodell 2012). The Canadian Association of Petroleum Producers claims that 175,000 wells have been fractured in Alberta and British Columbia without a single incident of harm to groundwater (Hussain 2012).

When anti-fracking activists challenge this blemish-free record – in Pennsylvania, the state Department of Environmental Protection fined Chesapeake Corp. almost US$1 million for contaminating 16 families' water wells with methane as a result of improper drilling practices – industry executives respond by adopting several strategies. First, they embrace an unacceptably narrow view of fracking, characterizing it as constituting just a single water blast. As Mooney (2011) points out, not only does this restricted definition overlook the fact that companies normally drill a dozen or more closely spaced wells at a single site, it also ignores the reality that fracking is a 'massive industrial endeavor that includes transporting, storing and pumping millions of gallons of water, and then managing almost equal volumes of flowback fluid'. If we adopt this broader definition, then shale gas drillers are far more likely to be guilty of some serious infractions.

Second, they argue that any leakage that may occur during the fracturing process is the result of contractor error, most notably faulty cementing and casing of wells. If done properly, the gap between the gas pipe and the wall of the well is filled with concrete, preventing any methane or chemical 'flowback' from seeping into the groundwater. Poor cementing, however, permits methane migration, leading to contamination. The iconic 2007 case of an explosion in a homeowner's basement in Bainbridge, Ohio seems to have been attributable to the faulty seal of a fracked well (Ehrenberg 2012). In other words, the pro-fracking forces contend that there is nothing necessarily wrong with the process that could not be remedied by improved industry practice. Critics challenge this interpretation. One alternative theory states that fracking may be creating some cracks that extend upward in the rock beyond the horizontal shale layer itself, linking up with other pre-existing fissures or openings from abandoned gas or oil wells. In this 'Swiss cheese of boreholes' methane moves up toward the surface, posing a groundwater risk (Mooney 2011).

Corporate interests have made themselves vulnerable in the battle for public opinion by concealing the composition of fracturing fluids. In the United States, the pro-fracking forces successfully inserted a provision in the 'Halliburton Loophole' that exempted companies from revealing the exact chemical recipe. They justified this on the basis of proprietary interest. Nevertheless, some of these agents, which include components of antifreeze and pesticides, are 'caustic, poisonous, or explosive' (Wilber 2012: 63).

Environmental conflicts are waged on symbolic as well as actual turf. In attempting to silence fractivists, industry and government forces have sometimes appeared heavy handed. For example, Lopi LaRoe, an Occupy Wall Street veteran and environmental activist, was recently served with a cease-and-desist order by the US Department of Agriculture's Forest Service division. LaRoe had begun to market merchandise (T-shirts, tote bags) with an image of conservation icon Smokey the Bear together with the slogan 'Only You Can Prevent Faucet Fires' (Rugh 2013).

Conclusion

In contrast to biodiversity loss, which has become firmly anchored on the global agenda, shale gas drilling or fracking is an environmental problem still largely under construction. To a great extent, this reflects a lack of scientific validation, especially in the early stages of issue assembly. More recently, public statements by Barrack Obama and David Cameron emphasizing the importance of shale gas production for their respective nations' energy security have made it considerably more difficult for anti-fracking forces to make significant headway in the political arena. Thus, in a recent opinion piece in the *New York Times*, Roger Cohen (2013) argues:

> Shale gas is an important new element in global energy supplies, a cost cutter and a geostrategic game changer that lessens Western dependence on Russia and Iran. But Cameron, by talking about the need for transparency while providing none, has undermined that future, and pretty Balcombe may well be precisely the kind of place least suited to it.

What this suggests is that the narrative here has shifted from the environmental and health risks of shale gas drilling to the need for public disclosure and debate before proceeding. If fracking is to avoid slipping into the dustbin of those environmental claims that failed to proceed through the final contestation stage of environmental problem construction, it needs a major breakthrough – a catastrophic event, an uncontestable medical finding or a sign-on by a formidable institutional sponsor or political champion.

Notes

1   The addition of yeast was permitted in a much later amendment.
2   In Australia, hydraulic fracturing is employed mainly to exploit coal-seam gas (CSG) rather than shale gas. According to a recent report in *The Economist*, CSG 'is transforming Australia's energy market and stimulating its robust economy' ('Gas goes boom' 2013).
3   Less well known than shale gas, shale oil is extracted using the same technology.
4   Ono and her late husband, John Lennon of the Beatles, purchased a farm in upstate New York near the Pennsylvania border atop the Marcellus Shale near the proposed site of a new natural gas pipeline.

Further reading

Wilber, T. (2012) *Under the Surface: Fracking, Fortunes, and the Fate of the Marcellus Shale*, Ithaca, New York: Cornell University Press.

Online resources

The Fracking Song ('My Water's On Fire Tonight') (YouTube video). Available HTTP: http://www.youtube.com/watch?v=timfvNgr_Q4
Waging Nonviolence: People-Powered News and Analysis. Available HTTP: http://wagingnonviolence.org

10 Conclusion

Environmental sociology is now half way through its fifth decade. The field first emerged in the 1970s, largely in the United States, as a reaction to growing social activism for environmental protection (Pretty *et al.* 2007: 2), something its founding figures – Fred Buttel, William Catton, Riley Dunlap, Alan Schnaiberg – didn't fully anticipate, but needed to explain. All subscribed, more or less, to the environmentalists' belief that humans are marching like lemmings on a path to societal ruin unless we radically change our ways. These first generation environmental sociologists attempted to unearth the causes of an approaching ecological crisis. Buttel (2003: 307) argues the best known concepts and theories in environmental sociology – Schnaiberg's 'treadmill of production'; Catton and Dunlap's dominant social paradigm/new ecological paradigm; Logan and Molotch's urban growth machine; Murphy's theory of the irrationality of capitalist–industrial rationality – were those that identified a key factor (or closely related set of factors) that had led to enduring an environmental crisis.

Rudel *et al.* (2011) distinguish between a first generation of theories about the political economy of the environment (impact treadmill, growth machine, and extraction theories) that attempted to show how the normal workings of industrial production damage the environment and a second wave of theories that focus on environmental destruction and the social movements that challenge the agents of destruction.[1] Rudel and his co-authors construe the latter as a 'double movement' in which environmental decline occurs but people mobilize to prevent further declines. This double movement resembles a pendulum, swinging first in the direction of environmental abuse; then, after a wave of protest occurs, towards significant regulation; and, finally, back towards the original start point, as free market forces gain traction and secure a significant rollback of regulations (Rudel *et al.* 2011: 226–7).

In its second stage of development, during the 1990s, environmental sociology shifted focus to finding a way out of the 'iron cage' of environmental despair; that is, discovering more effective mechanisms of environmental improvement and governance, rather than mainly explaining environmental degradation. Buttel

(2003) argues that environmental sociologists theorized and researched four strategies for environmental improvement: (1) mobilization of environmental movements, old and new; (2) sustaining or enhancing the environmental regulatory capacity of government; (3) ecological modernization, the notion that modern industrial societies can solve environmental problems through intensified development of innovative industrial technology, ecological efficiencies in production and consumption and green marketing (see Chapter 2 of this book) and (4) environmental internationalism via international environmental agreements, international environmental regimes and international intergovernmental organizations. Buttel concluded that environmental movements and activism are ultimately the most fundamental pillars of environmental reform. Citizen environmental mobilization, he argued, is the only force that can be counted on to challenge the continual backsliding and de-regulation that are characteristic of the 'business as usual' attitude of politicians and governments (Buttel 2003: 336).

During this period, the concepts of sustainability and sustainable development were positioned front and centre. Sustainable development, Sachs (1997: 71) observes, became the late twentieth-century expression for progress and social improvement (cited in Irwin 2001: 13). Following the lead of the Brundtland Commission Report, there was hope, if not certainty, that the economy and the environment were far more compatible than was previously thought.

Most recently, there has been yet another 'significant shift of problematics' to borrow Buttel's phrase. The notion that there is no inherent contradiction between sustainability and development has increasingly been called into question, while minimal attention has been given to 'the implications of rethinking sustainability for governance, security or ideas of justice' (Redclift 2009: 371–2). The environmental improvements that seemed to be on the cusp of possibility at the 1992 UN Conference on Environment and Development (Earth Summit) have generally stalled or proven to be unrealistic. The Kyoto Protocol is effectively, if not legally, dead. It seems doubtful that the Millennium Development goals will be met.

Meanwhile, a new 'doomsday on the horizon' (Willis 2013: 217), global climate change, has captured everyone's attention. Climate change is the heir to earlier catastrophic threats ranging from nuclear destruction to the collapse of economy and society predicted by the *Limits of Growth* (see Chapter 1). In response, there has been a rising chorus of voices demanding that sociology as a whole, and specifically environmental sociology, move the topic of global climate change to the epicentre of our discipline. The growing magnitude of climate changes, Rudel *et al.* (2011: 233) conclude, 'may alter the course of thinking about the political economy of the environment'.

In his plenary talk at the 2008 Annual Conference of the British Sociological Association, John Urry issued a 'call to arms' for sociologists, urging his audience to rectify its reluctance in the past to engage with the topic of climate change. This 'mixing of political urgency and need for profound rethinking of sociological traditions and conceptual frameworks' (Grundmann & Stehr 2010: 898) was manifest in an extended dialogue that played out on the pages of the journal *Current Sociology* between 2008 and 2010. In his introduction to the 2008 special issue entitled 'Two Dialogues: On Public Sociology and on Global Warming', editor Dennis Smith warned that sociology had a grave responsibility to bring its knowledge of how societies work and break down to the contest between greed (for profit) and fear (of climatic disaster), lest violent conflict break out everywhere. In her contribution to that symposium, Constance Lever-Tracy decried our disciplinary silence on the threat of global warming. 'We have already wasted too much time', she wrote, and 'may awaken too late to have any impact' (Lever-Tracy 2008a: 459).

Environmental sociologists have answered this call for action in various ways. Some have opted to revisit the task of unmasking those powerful social forces in capitalist–industrial civilization that underpin, and even promote, global destruction. In his introduction to a special issue of the *American Behavioral Scientist* devoted to 'Climate Change Skepticism and Denial', Riley Dunlap (2013) writes about 'pulling back the curtains' to reveal a 'disinformation' campaign waged by a loose coalition of industrial (especially fossil fuels) interests and conservative foundations and think tanks, often greatly aided by a small number of 'contrarian scientists' and by conservative media and politicians. This campaign is said to be the work of a 'denial machine' (Dunlap 2013), 'denier choir' (Elsasser & Dunlap 2013) or 'backlash coalition' (Lahsen 2013) funded by conservative foundations, corporations such as Exxon Mobil and associations such as the US Chamber of Commerce. Their message resonates in the 'conservative echo chamber', which consists of major conservative media, as well as a bevy of bloggers 'who work together to promote, one another, contrarian scientists, and all other components of the denial machine in a mutual effort to undermine the reality and threat of global warming' (Elsasser & Dunlap 2013: 755).

It isn't entirely clear who is guilty of orchestrating and conducting this satanic conservative choir, although it appears to be the same 'merchants of doubt' (Oreskes & Conway 2010) who in times past have downplayed and tried to dismiss the dangers of acid rain, cigarette smoking and herbicide use. What this conspiracy theorizing ignores is the position of many of the world's most powerful financial institutions – banks, global investors, insurance companies, the military – who have come to regard global climate change as the premier environmental threat today.[2] The World Bank, for example, is a major global funding source for initiatives in both climate change mitigation and adaptation. Their motivation for doing so may be more mercenary than moralistic – insurance companies, for example, stand to lose heavily if storms such as Hurricane Sandy become the norm. Nonetheless, it doesn't make sense to overlook their contribution to the global climate change debate and focus exclusively on the disinformation campaigns of energy companies and conservative propagandists.

Low carbon futures

More promising, I think, is the effort to develop a new disciplinary in sociology focusing on the prospects for and problems of low carbon futures. The noted British social theorist John Urry has been at the forefront of this effort. In two recent books, *Climate Change and Society* (2011) and *Societies Beyond Oil* (2013), Urry argues for a new 'resource turn' in sociology, whereby 'societies should be examined through the patterns, scale and character of their resource dependence and resource consequences' (Urry 2011: 16). The discipline, he argues, has long been 'carbon blind', hitching itself to a modern world where unlimited energy sources are taken for granted. Sociologists have long neglected vital issues pertaining to resource use, the potential for a catastrophic future and the necessity for the systematic restructuring of economic and social life. But, as we move closer to the advent of a new world dis/order, sociology can (and must) play an important role, displacing dominant economic models of human behaviour.

For Urry (2013) there are four possible futures for societies looking forward over the next three to four decades. The first such future is that a 'viable, global magic bullet' is discovered, replacing oil or at least making up for the energy shortages that occur once peak oil is reached. Urry has in mind a yet to be perfected alternative energy source such as hydrogen power, or a high-technology system such as geo-engineering. Inexplicably, he says nothing about the shale gas revolution (see Chapter 9 of this book). A second potential future, which he calls 'digital worlds', is not defined by discovering new energy supplies, but rather by innovating new technologies that demand less oil consumption. In particular, digital travel replaces the physical transport of people and things, much like virtual academic and corporate conferences do. The third future depicts a continuing dependence on fossil fuels, but the depletion of known sources leads to a new century of 'tough oil'. As the remaining 'oil dregs' become more difficult to extract, nations and corporations increasingly engage in intermittent resource wars. Urry dreads the emergence of a new world order, wherein the rich and powerful isolate themselves in fortified enclaves, while those who are not live in bleak, dystopian 'wild zones'. The fourth possible future (Urry himself prefers this one) sees a global transition to lower oil and other energy use. People here undertake 'an organized, planned powering down to low carbon lives' (Urry 2013: 158). Developing such a 'low carbon civil society', Urry stresses, 'is the pre-eminent global challenge to deal with the double whammy of rising temperatures and falling supplies of oil, which we cannot live with but will not be able to live without' (Urry 2013: 229).

Michael Redclift is another eminent sociologist who has addressed this issue of low carbon futures. In an article first published in *Current Sociology* (Redclift 2009) and subsequently reprised in the online journal *International Review of Social Science*, he argues that sociology has a real contribution to make on this important topic. Like Urry, Redclift believes that sociology (and particularly environmental sociology) is well positioned to assess the social dimensions of climate change and carbon capture (the processes through which economically developed societies have grown more dependent on carbon), and to detect possible routes out of this dependence. This quest should be heartening for sociology. It requires revisiting what we know, and subjecting environmental knowledge to new and unfamiliar investigations. It means investigating future alternatives to the 'hydrocarbon societies with which we are most familiar, just as Max Weber investigated unfamiliar whole societies in Antiquity' [Weber 1991] (Redclift 2009: 370).

Redclift suggests that one trump card held by sociology in its quest to become a leader in the search for alternatives to a carbon-based society is its fluency in treating 'whole societies' as utopias and imaginaries. The discipline has a long and honourable history, he observes, in serving as an acute lens through which to explore alternative ways of living and the way they correspond to, and connect with, wider human purposes (Redclift 2009: 370). In so doing, Redclift properly cautions, the political economy of the withdrawal from carbon dependence 'needs to be analyzed rather than evangelized'.

Not all researchers have shared the enthusiasm for studying low carbon futures indicated by Redclift and Urry. In his commentary on Redclift's (2011) article, Murphy (2011) predicts that it is likely that society will remain tied to the carbon economy for a long time. Rather than divert our attention to considering 'lofty post-carbon utopia discourse', he advises social scientists to intensify their efforts to expose and analyze the social causes and consequences of anthropogenic climate change. In particular, Murphy recommends undertaking a comparative study of the differences between Northern Europe and North America with regards to carbon dependence and commitment to achieving a more environmentally friendly, sustainable track.

In a book review symposium in the journal *Sociology*, Leon Sealey-Huggins, Amanda Rohloff and Mike Hulme (2012) all praise the 'kind of heightened sociological sensibility' about climate change demonstrated by Urry, but express some skepticism about his preferred strategy. Sealey-Huggins claims that Urry focuses too heavily on the concept of the post carbon, while overlooking the range of different politics surrounding climate change debates, especially those connected to environmental justice issues. Rohloff finds Urry's call for a shift to a market-regulated resource capitalism that is constrained by shrinking resources and increasing pollution doesn't go into enough detail to make it very useful. Hulme worries that Urry's depiction of possible energy futures is painted in tones that are too dark, gloomy and catastrophic. He recommends another narrative that is embedded in a humanistic tradition of virtue and hope.

Towards a new sociological narrative of energy futures

In carrying out such an analysis, we need 'the empirical richness and the theoretical creativity of sociology to conceptualize the post-carbon society' (Reusswig 2011:

189). In order to successfully decipher the transition to a post-carbon society, Reusswig argues, we need to think of this as a process of social change.

Additionally, we must apply our knowledge about past changes to better understand the future in a sober and holistic manner. As Lever-Tracy (2008b: 490) observes, 'Arguably, sociology is not qualified to evaluate the probability of the scenarios portrayed by natural scientists. What we can do is study the actual and likely responses in the context of power and culture'. I fully concur with this advice. We should be strongly encouraged to 'join the fray', that is, to engage with the academic and policy conversation on carbon dependency and a post carbon world. That said,

> If sociologists wish to join the conversation about carbon dependency, they have to be prepared to move beyond simplified narratives about evil carbon capitalists and heroic green crusaders. The political economy of energy is complex, and understanding it requires a firm grounding in environmental economics, international relations and political ecology.
>
> (Hannigan 2012b: 77–8)

While simplified narratives cannot be expected to provide much assistance here, this doesn't mean that we must at all costs shun an encompassing storyline. As Friedland and Mohr (2004: 38) observe,

> What a narrative does by definition is generate a story that subsumes the parts under the logic of the whole. . . . Cultural studies [also] tells stories, but ones that are not very good because of their theories' totalizing tendencies. Sociologists on the other hand no longer tell stories. We have not told a good story since the classical stories of disenchantment and commodification and their successors – secularization and modernization. Unlike imperialism, globalization as a story has zero content. Sociologists shy from collective purpose, avoid questions of 'why' in the pursuit of the 'how'.

Any such narrative must necessarily include the future of China. For the first time, China has overtaken the United States as the world's top oil importer. China is embarking on a vast programme of rapid urbanization: 250 million rural-to-urban migrants in a dozen years. In southern Shaanxi Province alone, 2.4 million farmers are being resettled from mountain areas to the towns and cities (Johnson 2013: 1). The central government has undertaken this massive population shift in order to create new markets for domestic products, rather than having to rely on exports. Does it make sense to interpret this through the lens of existing narratives of modernization, overshooting the limits of growth or cranking up the treadmills of production and consumption? Or should we be thinking about a new way of conceptualizing urban growth, energy consumption and environmental integrity?

Notes

1 Rudel's second wave of sociological theories about the political economy of the environment overlaps and bridges Buttel's two stages in the evolution of environmental sociology.

2 Ahead of the November, 2012 Doha round of negotiations on carbon emissions, a coalition of institutional investor groups from Europe, North America, Australia/ NewZealand, Asia and the United Nations, including three who manage US$20 trillion in assets, wrote an open letter stating that a rise in greenhouse gas emissions increases the risk of dangerous and disruptive climate change and urging world leaders to show decisive action over climate change and its related impacts (Blackburne 2012).

Bibliography

Abbey, E. (2006 [1975]) *The Monkey Wrench Gang*, New York: Harper Perennial.

Adam, B. (1996) 'Re-vision: the centrality of time for an ecological social science perspective', in S. Lash and B. Szerszynski (eds) *Risk, Environment and Modernity: Towards a New Ecology*, London: Sage.

Adams, W.M., Aveling, R., Brockington, D., Dickson, B., Elliott, J., Hutton, J., Roe, D., Vira, B. and Wolmer, W. (2004) 'Biodiversity conservation and the eradication of poverty', *Science*, 306(5699): 1146–9.

Agrawala, S., Broad, K. and Guston, D.H. (2001) 'Integrating climate forecasts and societal decision-making: challenges to an emergent boundary organization', *Science, Technology & Human Values*, 26(4): 454–77.

Alexander, J.C. and Smith, P. (1986) 'Social science and salvation: risk society as mythical discourse', *Zeitschrift für Soziologie*, 25: 251–62.

Alexandrescu, F.M. (2009) 'Not as natural as it seems: the social history of the environment in American sociology', *History of the Human Sciences*, 22(5): 47–80.

Alpert, P. (1993) 'Support for biodiversity research from the US Agency for International Development', *BioScience*, 43(9): 628–31.

Altheide, D. (1976) *Creating Reality: How TV News Distorts Events*, Beverly Hills, CA: Sage.

Anand, R. (2004) *International Environmental Justice: A North–South Dimension*, Aldershot: Ashgate.

Anderson, A. (1993a) 'Source–media relations: the production of the environmental agenda', in A. Hansen (ed.) *The Mass Media and Environmental Issues*, Leicester: Leicester University Press.

Anderson, A. (1993b) *Media, Culture and the Environment*, New Bruswick, NJ: Rutgers University Press.

Angier, N. (1994) 'Redefining diversity: biologists urge look beyond rain forests', *New York Times*, 29 November: B-5; B-9.

Anguelovski, I. and Roberts, D. (2011) 'Spatial justice and climate change: multiscale impacts and local development in Durban, South Africa', in J. Carmin and J. Agyeman (eds) *Environmental Inequalities Beyond Borders*, Cambridge, MA: The MIT Press.

Applebome, P. (2011) 'Drilling debate in village grows close and nasty', *New York Times*, 30 October: 1; 18.

Aronson, N. (1984) 'Science as a claims-making activity: implications for social problems research', in J. Schneider and J.I. Kitsuse (eds) *Studies in the Sociology of Social Problems*, Norwood, NJ: Ablex.

Arsel, M. and Büscher, B. (2012) 'Nature™ Inc.: changes and continuities in neoliberal conservation and market-based environmental policy', *Development and Change*, 43(1): 53–78.

Atkisson, A. (2011) *Believing Cassandra: How to be an Optimist in a Pessimist's World (2nd edition)*, London and Washington, DC: Earthscan.

Bachman, M. (2011) 'Fracking concerns arise in village', *YS (Yellow Springs) News*, 10 February. Online. Available HTTP: http://ysnews.com/news/2011/02/fracking-concerns-arise-in-village (accessed 3 February, 2014).

Bakhsh, N. (2013) 'Oil under green lawns'. Cited in 'Shale shock: the global impact of fracking', *National Post*, 5 July: FP9.

Balch, A. and Press, D. (2003) 'The politics of biodiversity: a political case of the endangered species act', in S.L. Spray and K.L. McGlothlin (eds) *Loss of Biodiversity*, Lanham, MD: Rowman & Littlefield.

Barkan, S. (1979) 'Strategic, tactical and organizational dilemmas of the protest movement against nuclear power', *Social Problems*, 27(1): 19–37.

Barry, J. (1999) *Environment and Social Theory*, London and New York: Routledge.

Barton, J.H. (1992) 'Biodiversity at Rio', *BioScience*, 42(10): 773–6.

Baumann, E.A. (1989) 'Research rhetoric and the social construction of elder abuse', in J. Best (ed.) *Images of Issues: Typifying Contemporary Social Problems*, New York: Aldine de Gruyter.

BBC News Africa (2013) 'Goldman prize for South African anti-fracking warrior', 15 April, Online. Available HTTP: http://www.bbc.co.uk/news/world-africa-22151746 (accessed 25 August, 2013).

Beck, U. (1992) *The Risk Society: Towards a New Modernity*, London: Sage.

Beck, U. and van Loon, J. (2011) 'Until the last ton of fossil fuel has burnt to ashes: climate change, global inequalities and the dilemma of green politics', in D. Held, A. Hervey and M. Theros (eds) *The Governance of Climate Change*, Cambridge: Polity Press.

Begos, K. (2013) 'Fracking critics unhappy with Obama climate speech', St. Louis Post-Dispatch, 27 June. Online. Available HTTP: http://www.stltoday.com/business/local/fracking-critics-unhappy-with-obama-climate-speech/article_ca856686-7fd0-5f98-a6cd-f589dca4d6d0.html (accessed 5 March, 2014).

Begos, K. and Peltz, J. (2013) 'Celebrity fractivists: true advocates or NIMBYs', *Salon*, 5 March. Online. Available HTTP: http://www.salon.com/2013/03/05/celebrity_fractivists_true_advocates_or_nimbys/ (accessed 25 August, 2013).

Benedick, R.E. (1991) *Ozone Diplomacy: New Dimensions in Safeguarding the Planet*, Cambridge, MA and London: Harvard University Press.

Benford, R.D. (1993) 'Frame disputes within the nuclear disarmament movement', *Social Forces*, 71(3): 677–701.

Benton, T. (1991) 'Biology and social science: why the return of the repressed should be given a (cautious) welcome', *Sociology*, 25(1): 1–29.

Berg, C. (2013) 'Collaborative group leads city environmental clean-up drive', *Chinadaily USA*, 24 June. Online. Available HTTP: http://usa.chinadaily.com.cn/epaper/2013-06/24/content_16650407.htm (accessed 25 August, 2013).

Best, J. (1987) 'Rhetoric in claims-making', *Social Problems*, 34(2): 101–21.

Best, J. (1989) 'Afterword: extending the constructionist perspective: a conclusion – and an introduction', in J. Best (ed.) *Images of Issues: Typifying Contemporary Social Problems*, New York: Aldine de Gruyter.

Betts, K. (1997) 'Review of J. Hannigan, *Environmental Sociology: A Social Constructionist Perspective* [London and New York: Routledge, 1995]', *The Australian and New Zealand Journal of Sociology*, 33(1): 103–5.

Bierstedt, R. (1981) *American Sociological Theory: A Critical History*, New York: Academic Press.

Blackburne, A. (2012) 'Investors demand government action to avoid dangerous climate change impacts', *blue&greentomorrow*, 21 November. Online. Available HTTP: http://blueandgreentomorrow.com/2012/11/21/investors-demand-government-action-to-avoid-dangerous-climate-change-impacts/ (accessed 25 August, 2013).

Blakeslee, A.M. (1994) 'The rhetorical construction of novelty: presenting claims in a letters forum', *Science, Technology & Human Values*, 19(1): 88–100.

Bloomfield, B.P. and Doolin, B. (2013) 'Symbolic communication in public protest over genetic modification: visual rhetoric, symbolic excess, and social mores', *Science Communication*, 35(4): 502–27.

Blowers, A. (1993) 'Environmental policy: the quest for sustainable development', *Urban Studies*, 30(4/5): 775–96.

Blowers, A. (1997) 'Environmental policy: ecological modernization or the risk society?' *Urban Studies*, 34(5/6): 845–71.

Blowers, A., Lowry, D. and Solomon, B.D. (1991) *The International Politics of Nuclear Waste*, London: Macmillan.

Blühdorn, I. (2000) 'Ecological modernization and post-ecologist politics', in G. Spaargaren, A.P.J. Mol and F. Buttel (eds) *Environment and Global Modernity*, London: Sage.

Boardman, R. (1991) *International Organization and the Conservation of Nature*, London: Macmillan.

Bocking, S. (1993) 'Conserving nature and building a science: British ecologists and the origins of the Nature Conservancy', in M. Shortland (ed.) *Science and Nature: Essays in the History of the Environmental Sciences*, Oxford: British Society for the History of Science.

Boehmer-Christiansen, S. (2003) 'Science, equity and the war against carbon', *Science, Technology & Human Values*, 28(1): 69–92.

Bogan, J. (2009) 'The father of shale gas', *Forbes*, 7 July. Online. Available HTTP: http://www.forbes.com/2009/07/16/george-mitchell-gas-business-energy-shale.html (accessed 30 October, 2013).

Boia, L. (2005) *The Weather in the Imagination*, London: Reaktion Books.

Bora, A. and Dobert, R. (1992) 'Prerequisites of procedural rationality: notes on a German technology assessment of transgenic plants with herbicide resistance', paper presented at the Symposium, 'Current Developments in Environmental Sociology', International Sociological Association, Woudschoten, the Netherlands.

Bosso, C.J. (1987) *Pesticides and Politics: The Life Cycle of a Public Issue*, Pittsburgh, PA: University of Pittsburgh Press.

Boulding, K.E. (1950) *A Reconstruction of Economics*, New York: Wiley.

Boulding, K. (1966) 'The economics of coming spaceship Earth', in H. Jarett (ed.) *Environmental Quality in a Growing Economy*, Baltimore, MD: Resources for the Future, Johns Hopkins University Press.

Boykoff, M.T. (2011) *Who Speaks for the Climate? Making Sense of Media Reporting on Climate Change*, Cambridge: Cambridge University Press.

Boyne, R. (2003) *Risk*, Buckingham and Philadelphia, PA: Open University Press.

Bramble, B.J. and Parker, G. (1992) 'Non-governmental organizations and the making of U.S. international environmental policy', in A. Hurrell and B. Kingsbury (eds) *The International Politics of the Environment*, Oxford: Clarendon Press.

Bramwell, A. (1989) *Ecology in the 20th Century: A History*, New Haven and London: Yale University Press.

Braun, B. and Castree, N. (1998) *Remaking Reality: Nature at the Millennium*, London and New York: Routledge.

Bray, D. (2010) 'The scientific consensus of climate change revisited', *Environmental Science & Policy*, 13(5): 340–50.

Breyman, S. (1993) 'Knowledge as power: ecology movements and global environmental problems', in R.D. Lipschutz and K. Conca (eds) *The State and Social Power in Global Environmental Politics*, New York: Columbia University Press.

Brookes, S.H., Jordan, A.G., Kimber, R.H. and Richardson, J.J. (1976) 'The growth of the environment as a political issue in Britain', *British Journal of Political Science*, 6(2): 245–55.

Brown, E., Hartman, K., Borick, C., Rabe, B.G. and Ivacko, T. (2013) *Public Opinions on Fracking: Perspectives from Michigan and Pennsylvania*, The Center for Local, State, and Urban Policy, Gerald R. Ford School of Public Policy, University of Michigan, May.

Brown, J. (2013) 'Fracking sticks in your mind because it is such a horrible word. Malton has mixed feelings about its new prospects'. *The Independent*, 27 June. Online. Available HTTP: http://www.independent.co.uk/news/uk/politics/fracking-sticks-in-your-mind-because-it-is-such-a-horrible-word-malton-has-mixed-feelings-about-its-new-prospects-8677591.html (accessed 30 October, 2013).

Brown, L.R. (1986) 'And today we're going to talk about biodiversity . . . that's right, biodiversity', in E.O. Wilson (ed.) *Biodiversity*, Washington, DC: National Academy Press.

Brown, L.R. (2000) 'Overview: the acceleration of change', in L. Brown, M. Renner and B. Halwell (eds), *Vital Signs 2000*, New York: W.W. Norton & Company.

Brown, L.R. (2011) *World on the Edge: How to Prevent Environmental and Economic Collapse*, New York: Earth Policy Institute/W.W. Norton & Company.

Brown, L.R., Renner, M., Halwell, B., with Abramovitz, J.N. (*et al.*) (2000) *Vital Signs 2000: The Environmental Trends That Are Shaping Our Future*, New York: Norton.

Brown, N. and Michael, M. (2004) 'Risky creatures: institutional species boundary change in biotechnology regulation', *Health, Risk & Society*, 6(3): 207–22.

Browne, P. (2010) 'Tapping Methane at Lake Kivu in Africa', *The New York Times*, 26 January. Online. Available HTTP: http://green.blogs.nytimes.com/2010/01/26/tapping-methane-at-africas-lake-kivu/?_r=0 (accessed 30 October, 2013).

Brulle, R.J. (1998) 'Review of Hannigan (1995)', *Sociological Inquiry*, 68(1): 137–9.

Brulle, R.J. (2000) *Agency, Democracy and Nature: The US Environmental Movement from a Critical Theory Perspective*, Cambridge, MA and London: The MIT Press.

Bryant, B. and Mohai, P. (1992) 'Introduction', in B. Bryant and P. Mohai (eds) *Race and the Incidence of Environmental Hazards: A Time for Discourse*, Boulder, CO: Westview Press.

Bryant, R. (2010) 'Peering into the abyss: environment, research, and absurdity in the "age of stupid"', in M.R. Redclift and G. Woodgate (eds), *The International Handbook of Environmental Sociology (2nd edition)*, Cheltenham: Edward Elgar.

Budiansky, S. (1995) *Nature's Keepers: The New Science of Nature Management*, New York: Free Press.

Buell, F. (2003) *From Apocalypse to Way of Life: Environmental Crisis in the American Century*, London and New York: Routledge.

Bullard, R.D. (1990) *Dumping in Dixie: Race, Class and Environmental Quality*, Boulder, CO: Westview Press.

Burgess, A. (2003) 'Review of P. Harremoes, D. Gee, M. MacGarvin, A. Stirling, J. Keys, B. Wynne and S. Guedes Vaz, *The Precautionary Principle in the 20th Century: Late Lessons from Early Warnings* (London: Earthscan, 2002)', *Health, Risk & Society*, 5(1): 105–7.

Burgess, J. and Harrison, C.M. (1993) 'The circulation of claims in the cultural politics of environmental change', in A. Hansen (ed.) *The Mass Media and Environmental Issues*, Leicester: Leicester University Press.

Burns, T.J. and LeMoyne, T. (2001) 'How environmental movements can be more effective: prioritizing environmental themes in political discourse', *Human Ecology Review*, 8(1): 26–38.

Burros, M. (2005) 'Stores say wild salmon, but tests say farm bred', *The New York Times*, 10 April.

Burton, A. (2003) 'Extinction of taxonomists hinders conservation', *Frontiers in Ecology*, 1(5): 231.

Büscher, B. (2010) 'Anti-politics as political strategy: neoliberalism and transfrontier conservation in Southern Africa', *Development and Change*, 41(1): 29–51.

Büscher, B. (2013a) 'Transforming the frontier: peace parks and the politics of neoliberal conservation in Southern Africa', presented at the Critical Border Studies Speaker Series, Center for International Security Studies, York University/Centre for Diaspora and Transnational Studies, University of Toronto, 17 May.

Büscher, B. (2013b) *Transforming the Frontier: Peace Parks and the Politics of Neoliberal Conservation in Southern Africa*, Durham, NC: Duke University Press.

Büscher, B. and Whande, W. (2007) 'Whims of the wind: emerging trends in biodiversity conservation and protected area management,' *Conservation & Society*, 5(1): 22–43.

Butigan, K. (2013) 'Sandra Steingraber's latest experiment for a fracking-free world,' *Waging Nonviolence*, 25 April. Online. Available HTTP: http://wagingnonviolence.org/feature/sandra-steingrabers-latest-experiment-for-a-fracking-free-world/ (accessed 26 August, 2013).

Buttel, F.H. (1986) 'Sociology and the environment: the winding road toward human ecology', *International Social Science Journal*, 38(3): 337–56.

Buttel, F.H. (1987) 'New directions in environmental sociology', *Annual Review of Sociology*, 13: 465–88.

Buttel, F.H. (2000) 'Classical theory and contemporary environmental sociology: some reflections on the antecedents and prospects for reflexive modernization theories in the study of environment and society', in G. Spaargaren, A.P.J. Mol and F.H. Buttel (eds) *Environment and Global Modernity*, London: Sage.

Buttel, F.H. (2002a) 'Environmental sociology and the classical sociological tradition: some observations on current controversies', in R.E. Dunlap, F.H. Buttel, P. Dickens and A. Gijswijt (eds) *Sociological Theory and the Environment: Classical Foundations, Contemporary Insights*, Lanham, MD: Rowman & Littlefield.

Buttel, F.H. (2002b) 'Has environmental sociology arrived?' *Organization & Environment*, 15(1): 42–54.

Buttel, F.H. (2003) 'Environmental sociology and the explanation of environmental reform,' *Organization & Environment*, 16(3): 306–44.

Buttel, F.H. (2004) 'The treadmill of production: an appreciation, assessment, and agenda for research', *Organization & Environment*, 17(3): 323–36.

Buttel, F.H. and Taylor, P. (1992) 'Environmental sociology and global environmental change: a critical assessment', *Society and Natural Resources*, 5(3): 211–30.

Buttel, F.H., Dickens, P., Dunlap, R.E. and Gijswijt, A. (2002) 'Sociological theory and the environment: an overview and introduction', in R.E. Dunlap, F.H. Buttel, P. Dickens and A. Gijswijt (eds) *Sociological Theory and the Environment: Classical Foundations, Contemporary Insights*, Lanham, MD: Rowman & Littlefield.

Cable, S. and Benson, M. (1993) 'Acting locally: environmental injustice and the emergence of grass-roots environmental organizations', *Social Problems*, 40(4): 464.

Cable, S. and Cable, C. (1995) *Environmental Problems, Grassroots Solutions: The Politics of Grassroots Environmental Conflict*, New York: St. Martin's Press.

Calhoun, C. (2004) 'A world of emergencies: fear, intervention, and the limits of cosmopolitan order', *Canadian Review of Sociology and Anthropology*, 41(4): 373–95.

Cantrill, J.G. (1992) 'Understanding environmental advocacy: interdisciplinary research and the role of cognition', *The Journal of Environmental Education*, 24(1): 35–42.

Capek, S.M. (1993) 'The "environmental justice" frame: a conceptual discussion and an application', *Social Problems*, 40(1): 5–24.

Capuzza, J. (1992) 'A critical analysis of image management within the environmental movement', *The Journal of Environmental Education*, 24(1): 9–14.

Carlton, J. (2005) 'Green groups, ranchers bury hatchet to curb oil drilling', *The Wall Street Journal*, 23 March: B1, B4.

Carolan, M.S. (2006) 'The values and vulnerabilities of metaphors within the environmental sciences,' *Society & Natural Resources*, 19(10): 921–30.

Carolan, M.S. and Bell, M.M. (2004) 'No fence can stop it: debating dioxin drift from a small U.S. town to Arctic Canada', in N.E. Harrison and G.C. Bryner (eds) *Science and Politics in the International Environment*, Lanham, MD: Rowman & Littlefield.

Carroll, J. and Polson, J. (2012) 'Fracking land bubble fear,' *National Post*, 10 January: FP 12.

Carson, R. (1962/2002) *Silent Spring*, Boston: Houghton Mifflin.

Catton, W.R. Jr. (1980) *Overshoot: The Ecological Basis of Revolutionary Change*, Urbana, IL: University of Illinois Press.

Catton, W.R. Jr. (1994) 'Foundations of human ecology', *Sociological Perspectives*, 37(1): 75–95.

Catton, W.R. Jr. (2002) 'Has the Durkheim legacy misled sociology?', in R.E. Dunlap, F.H. Buttel, P. Dickens and A. Gijswijt (eds) *Sociological Theory and the Environment: Classical Foundations, Contemporary Insights*, Lanham, MD: Rowman & Littlefield.

Catton, W.R. Jr. and Dunlap, R.E. (1978) 'Environmental sociology: a new paradigm', *The American Sociologist*, 13: 41–9.

Chavez, C. (1993) 'Farm workers at risk', in R. Hofrichter (ed.) *Toxic Struggles: The Theory and Practice of Environmental Justice*, Philadelphia, PA: New Society Publishers.

Christoforou, T. (2003) 'The precautionary principle and democratizing expertise: a European legal perspective', *Science and Public Policy*, 30(3): 205–11.

Churchill, W. and LaDuke, W. (1985) 'Radioactive colonization and the Native American', *Socialist Review*, 15: 95–120.

Ciccantell, P.S. (1999) 'It's all about power: the political economy and ecology of redefining the Brazilian Amazon', *The Sociological Quarterly*, 40(2): 293–315.

Cittadino, E. (1993) 'The failed promise of human ecology', in M. Shortland (ed.) *Science and Nature: Essays in the History of the Environmental Sciences*, Oxford: British Society for the History of Science.

Clapp, J. and Dauvergne, P. (2005) *Paths to a Green World: The Political Economy of the Global Environment*, Cambridge, MA and London: The MIT Press.

Clarke, D. (1981) 'Second-hand news: production and reproduction at a major Ontario television station', in L. Salter (ed.) *Communication Studies in Canada*, Toronto: Butterworth.

Clarke, D. (1992) 'Constraints of television news production: the example of story geography', in M. Grenier (ed.) *Critical Studies in Canadian Mass Media*, Toronto: Butterworth.

Clarke, L. (1988) 'Explaining choices among technological risks', *Social Problems*, 35(1): 22–35.

Clarke, L. and Short, J.F. Jr. (1993) 'Social organization and risk: some current controversies', *Annual Review of Sociology*, 19: 375–99.

Clements, F.E. (1905) *Research Methods in Ecology*, Lincoln, NB: University Publishing Company. Reprinted by Arno Press (1977).

Cline, W.R. (1992) *The Economics of Global Warming*, Washington, DC: Institute for International Economics.

Cohen, J.L. (1985) 'Strategy or identity: new theoretical paradigms and contemporary social movements', *Social Research*, 52(4): 663–716.

Cohen, R. (2013) 'Britain's furour over fracking', *New York Times*, 26 August. Online. Available HTTP: http://www.nytimes.com/2013/08/27/opinion/roger-cohen-britains-furor-over-fracking.html?_r=0 (accessed 30 October, 2013).

Collingridge, D. and Reeve, C. (1986) *Science Speaks to Power: The Role of Experts in Policy Making*, London: Frances Pinter.

Commoner, B. (1971) *The Closing Circle*, New York: Bantam Books.

Conklin, B. and Graham, L. (1995) 'The shifting middle ground: Amazonian Indians and eco-politics', *American Anthropologist*, 97(4): 695–710.

Cooley, C.H. (1902) *Human Nature and the Social Order*, New York: Scribner's.

Corbett, J.B. (1993) 'Atmospheric ozone: a global or local issue? Coverage in Canadian and U.S. newspapers', *Canadian Journal of Communication*, 18(1): 81–7.

Cormier, J. and Tindall, D.B. (2005) 'Wood frames: framing the forests in British Columbia', *Sociological Focus*, 38(1): 1–24.

Corner, J. and Richardson, K. (1993) 'Environmental communication and the contingency of meaning: a research note', in A. Hansen (ed.) *The Mass Media and Environmental Issues*, Leicester: Leicester University Press.

Cotgrove, S. (1991) 'Sociology and the environment: Cotgrove replies to Newby', *Network* (British Sociological Association), 51, October.

Cottle, S. (1993) 'Mediating the environment: modalities of TV news', in A. Hansen (ed.) *The Mass Media and Environmental Issues*, Leicester: Leicester University Press.

Covello, V.T and Johnson, B.B. (1987) 'The social and cultural construction of risk: issues, methods and case studies', in B.B. Johnson and V.T. Covello (eds) *The Social and Cultural Contruction of Risk Selection and Perception*, Dordrecht, Holland: D. Reidel Publishing Company.

Cowan, J. (2008) 'A pipe dream: Rwanda to tap methane gas from lake in a risky project', *National Post* (Toronto), 10 June: A3.

Cowling, E.B. (1982) 'Acid precipitation in historical perspective', *Environmental Science and Technology*, 16(2): 110A–123A.

Cracknell, J. (1993) 'Issues arenas, pressure groups and environmental agendas', in A. Hansen (ed.) *The Mass Media and Environmental Issues*, Leicester: Leicester University Press.

Crist, E. (2004) 'Against the social construction of nature and wilderness', *Environmental Ethics*, 26(1): 5–24.

Cronon, W. (1996) 'The trouble with wilderness; or getting back to the wrong nature', *Environmental History*, 1(1): 7–28.

Cronor, C.F. (1993) *Regulating Toxic Substances: A Philosophy of Science and Law*, New York: Oxford University Press.

Curtin, P.A. and Rhodenbaugh, E. (2001) 'Building the news media agenda on the environment: a comparison of public relations and journalistic sources,' *Public Relations Review*, 27(2): 179–95.

Cylke, F.K. Jr. (1993) *The Environment*, New York: HarperCollins College Publishers.

Dahl, R. (1961) *James and the Giant Peach*. Illustrated by Nancy Ekolm Burkert. New York: Alfred A. Knopf, Inc.

Dake, K. (1992) 'Myths of nature: culture and social construction of risk', *Journal of Social Issues*, 48(4): 21–37.

Daley, P. and O'Neil, D. (1991) 'Sad is too mild a word: press coverage of the *Exxon Valdez* Oil Spill', *Journal of Communication*, 41(4): 42–57.

Davidson, D.J. and Frickel, S. (2004) 'Understanding environmental governance: a critical review', *Organization & Environment*, 17(4): 471–92.

Davidson, D.J. and Gismondi, M. (2011) *Challenging Legitimacy at the Precipice of Energy Calamity*, New York: Springer.

Davidson, D.J. and MacKendrick, N. (2004) 'All dressed up and nowhere to go: the discourse of ecological modernization', *The Canadian Review of Sociology and Anthropology*, 4(1): 47–65.

Davis, D.H. (2007) *Ignoring the Apocalypse. Why Planning to Prevent Environmental Catastrophe Goes Astray*, Westport, CT: Praeger.

Desfor, G. and Keil, R. (2004) *Nature and the City: Making Environmental Policy in Toronto and Los Angeles*, Tuscon, AZ: The University of Arizona Press.

Dewar, E. (1995) *Cloak of Green*, Toronto: James Lorimer & Co.

Diamond, J. (2005) *Collapse: How Societies Choose to Fail or Succeed*, New York: Viking.

Diamond, J. and Ordunia, D. (1997) *Guns, Germs and Steel*, New York: Norton.

Dickens, P. (1992) *Society and Nature: Towards a Greener Social Theory*, Hemel Hempstead: Harvester Wheatsheaf.

Dietz, T.R. and Rycroft, R.W. (1987) *The Risk Professionals*, New York: Russell Sage Foundation.

Dietz, T.R. *et al.* (2002) 'Risk, technology and society', in R.E. Dunlap and W. Michelson (eds) (2002) *Handbook of Environmental Sociology*, Westport, CT: Greenwood Press.

Dispensa, J.M. and Brulle, R.J. (2003) 'Media's social construction of environmental issues: focus on global warming – a comparative study', *International Journal of Environmental Issues*, 23(10): 74–105.

Dorfman, A. (1992) 'Sideshows galore', *Time*, 139(22), 1 June: 43.

Doughty, R.W. (1975) *Feather Fashions and Bird Preservation: A Study in Nature Protection*, Berkeley and Los Angeles, CA: University of California Press.

Douglas, M.A. and Wildavsky, A. (1982) *Risk and Culture: An Essay on the Selection of Technological and Environmental Dangers*, Berkeley, CA: University of California Press.

Dowling, J. and Pfeffer, J. (1975) 'Organizational legitimacy: social values and organizational behavior', *Pacific Sociological Review*, 18(1): 122–36.

Downs, A. (1972) 'Up and down with ecology – the "issue-attention" cycle', *The Public Interest*, 28: 38–50.

Doyle, J. (1985) *Altered Harvest: Agriculture, Genetics and the Fate of the World's Food Supply*, New York: Penguin.

Dryzek, J.S. (1983) 'Ecological rationality', *International Journal of Environmental Studies* 21(1): 5–10.

Dryzek, J.S. (1997; 2nd edition 2005) *The Politics of the Earth: Environmental Discourses (2nd edition)*, Oxford and New York: Oxford University Press.

Dumanowski, D. (1994) 'Mudslinging on the earth beat,' *Amicus Journal*, 15 (Winter): 40–1.

Dumoulin, D. (2003) 'Local knowledge in the hands of transnational networks: a Mexican viewpoint', *International Social Science Journal*, 178 (December): 593–605.

Duncan, O.D. (1961) 'From social system to ecosystem', *Sociological Inquiry*, 31(2): 140–9.

Dunk, T. (1994) 'Talking about trees: environment and society in forest workers' culture', *Canadian Review of Sociology and Anthropology* 31(1): 14–34.

Dunlap, R.E. (1993) 'From environmental problems to ecological problems', in C. Calhoun and G. Ritzer (eds), *Social Problems*, New York: McGraw Hill.

Dunlap, R.E. (2002a) 'Paradigms, theories and environmental sociology', in R.E. Dunlap, F.H. Buttel, P. Dickens and A. Gijswijt (eds) *Sociological Theory and the Environment: Classical Foundations, Contemporary Insights*, Lanham, MD: Rowman & Littlefield.

Dunlap, R.E. (2002b) 'Environmental sociology: a personal perspective on its first quarter century', *Organization & Environment*, 15(1): 10–29.

Dunlap, R.E. (2010) 'The maturation and diversification of environmental sociology: from constructivism and realism to agnosticism and pragmatism', in M.R. Redclift and G. Woodgate (eds), *International Handbook of Environmental Sociology (2nd edition)*, Cheltenham: Edward Elgar.

Dunlap, R.E. (2013) 'Climate change and denial: an introduction', *American Behavioral Scientist*, 57(6): 691–8.

Dunlap, R.E. and Catton, W.R. Jr. (1979) 'Environmental sociology', *Annual Review of Sociology*, 5(1): 243–73.

Dunlap, R.E. and Catton, W.R. Jr. (1992/3) 'Towards an ecological sociology: the development, current status and probable future of environmental sociology', *Annals of the International Institute of Sociology*, 3 (New Series): 263–84.

Dunlap, R.E. and Michelson, W. (eds) (2002) *Handbook of Environmental Sociology*, Westport, CT: Greenwood Press.

Dunlap, R.E., Buttel, F.H., Dickens, P. and Gijswijt, A. (eds) (2002) *Sociological Theory and the Environment: Classical Foundations, Contemporary Insights*, Lanham, MD: Rowman & Littlefield.

Dunlap, R.E., Xiao, C. and McCright, A.M. (2001) 'Politics and environment in America: partisan and ideological cleavages in public support for environmentalism', *Environmental Politics*, 10(4): 23–48.

Dunlap, T.R. (1991) 'Organisation and wildlife preservation: the case of the whooping crane in North America', *Social Studies of Science*, 21(2): 197–221.

Dunwoody, S. and Griffin, R.L. (1993) 'Journalistic strategies for reporting long-term environmental issues: a case study of three Superfund sites', in A. Hansen (ed.) *The Mass Media and Environmental Issues*, Leicester: Leicester University Press.

Durkheim, E., with an introduction by Lewis A. Coser; translated by W.D. Halls (1984 [1893]) *The Division of Labor in Society*, New York: Free Press.

Durkheim, E. (2002) [1895] 'The rules of sociological method', in C. Calhoun, J. Gerteis, J. Moody, S. Pfaff, K. Schmidt and I. Virk (eds) *Classical Sociological Theory*, Malden, MA and Oxford: Blackwell.

Economist, The (2012) 'Sorting frack from fiction', 14 July.

Edelman, M. (1964) *The Symbolic Use of Politics*, Urbana, IL: University of Illinois Press.

Edelman, M. (1977) *Political Language: Words That Succeed and Policies That Fail*, New York: Academic Press.

Eden, S., Tunstall, S.M. and Tapsell, S.M. (2000) 'Translating nature: river restoration as nature-culture', *Environment and Planning D: Society and Space*, 18(2): 257–73.

Eder, K. (1996) *The Social Construction of Nature: A Sociology of Ecological Enlightenment*, London: Sage.

Ehrenberg, R. (2012) 'The facts behind the frack', *Science News*, 182(5): 8.

Ehrlich, P.R. (1968) *The Population Bomb*, New York: Ballantine Books.

Ehrlich, P.R. and Ehrlich, A. (1981) *Extinction: The Causes and Consequences of the Disappearance of Species*, New York: Random House.

Einsiedal, E. and Coughlan, E. (1993) 'The Canadian press and the environment: reconstructing a social reality', in A. Hansen (ed.) *The Mass Media and Environmental Issues*, Leicester: Leicester University Press.

Eisner, T. (1989–90) 'Prospecting for nature's chemical riches', *Issues in Science and Technology*, 6(2): 31–4.

Eisner, T. and Beiring, E.A. (1994) 'Biotic Exploration Fund: protecting biodiversity through chemical prospecting', *BioScience*, 44(2): 95–8.

Eldredge, N. (1991) *The Miner's Canary: Extinctions Past and Present*, New York: Prentice Hall.

Eldredge, N. (1992–3) 'Confronting the skeptics' (Review of E.O. Wilson, *The Diversity of Life* [Cambridge, MA: Harvard University Press, 1992]', *Issues in Science and Technology*, 9(2): 90–2.

Elliott, A. (2002) 'Beck's sociology of risk: a critical assessment', *Sociology* 36(2): 293–315.

Elsasser, S.W. and Dunlap, R.E. (2013) 'Leading voices in the denier choir: conservative columnists' dismissal of global warming and denigration of climate science', *American Behavioral Scientist*, 57(6): 754–76.

'End near? Doomsday Clock holds at 5 'til midnight' (2013) *Yahoo News*, 15 January. Online. Available HTTP: http://news.yahoo.com/end-near-doomsday-clock-holds-5-til-midnight-232147095.html (accessed 26 August, 2013).

Enloe, C.H. (1975) *The Politics of Pollution in a Comparative Perspective: Ecology and Power in Four Nations*, New York: David McKay Co. Inc.

'Environmentalists mark milestone "war in the woods" protest' (2013) *Metro* (Toronto), 12 August: 8.

Environmental Conservation (1993) 'Editor's note: the biodiversity treaty', *Environmental Conservation*, 20(3): 277.

Enzenberger, H.M. (1979) 'A critique of political ecology', in A. Cockburn and J. Ridgeway (eds) *Political Ecology*, New York: Quadrangle.

Escobar, A. (1996) 'Constructing nature: elements for a post-structural political ecology', in R. Peet and M. Watts (eds) *Liberation Ecology*, London: Routledge.

Etzioni, A. (1961) *A Comparative Analysis of Complex Organizations: On Power, Involvement and Their Correlates*, Glencoe, IL: The Free Press.

European Environment Agency (2012) *Consumption and the Environment – 2012 Update*, Copenhagen: European Environment Agency.

Evans, D. (1992) *A History of Nature Conservation in Britain*, London and New York: Routledge.

Evans, P. (2002) 'Introduction: looking for agents of urban livability in a globalized political economy', in P. Evans (ed.) *Livable Cities? Urban Struggles for Livelihood and Sustainability*, Berkeley and Los Angeles, CA: University of California Press.

Eyerman, R. and Jamison, A. (1989) 'Environmental knowledge as an organizational weapon: the case of Greenpeace', *Social Science Information*, 28(1): 99–119.

Faber, D. and O'Connor, J. (1993) 'Capitalism and the crisis of environmentalism', in R. Hofrichter (ed.) *Toxic Struggles: The Theory and Practice of Environmental Justice*, Philadelphia, PA: New Society Publishers.

Faro, A.L. (2004) 'Actors, conflicts and the globalization movement', *Current Sociology*, 52(4): 633–47.

Ferree, M.M., Gamson, W.A., Gerhards, J. and Rucht, D. (2002) *Shaping Abortion Discourse: Democracy and the Public Sphere in Germany and the United States*, Cambridge: Cambridge University Press.

Fico, D.E., Li, X. and Kim, Y. (2000) 'Changing work environment of environmental reporters', *Newspaper Research Journal*, 21 (Winter): 2–12.

Fisher, D.R. (2003) 'Global and domestic actors within the global climate change regime: toward a theory of the global environmental system', *International Journal of Sociology and Social Policy*, 23(10): 5–30.

Fishman, M. (1980) *Manufacturing the News*, Austin, TX: University of Texas Press.

Flanagan, E. (2012) 'Fighting fracking in South Africa and beyond', *Waging Nonviolence*, 10 September. Online. Available HTTP: http://wagingnonviolence.org/feature/fighting-fracking-in-south-africa-and-beyond/ (accessed August 27, 2013).

Fletcher, F.J. and Stahlbrand, L. (1992) 'Mirror or participant? The news media and environmental policy', in R. Boardman (ed.) *Canadian Environmental Policy: Ecosystems, Politics and Process*, Toronto: Oxford University Press.

Foster, J.B. (1999) 'Marx's theory of metabolic rift: classical foundations for an environmental sociology', *American Journal of Sociology*, 105(2): 366–405.

Foster, J.B. and Holleman, H. (2012) 'Weber and the environment: classical foundations for a postexemptionalist sociology', *American Journal of Sociology*, 117(6): 1625–73.

Foster, P. (2012) '2052? More like 2084', *National Post* (Toronto), 15 June: FP15.

Foucault, M. (1979) *Discipline and Punish: The Birth of the Prison*, trans. A. Sheridan, New York: Vintage.

Foucault, M. (1980) *Power/knowledge. Selected Interviews and Other Writings 1972–1977* (ed. C. Gordon), Brighton: Harvester.

Fox, S.R. (1981) *John Muir and His Legacy: The American Conservation Movement*, Boston: Little, Brown & Co.

'Fracking: energy futures' (2013) *The Guardian*, 27 June.

'Fracking for oil causes earth to shake, rattle and roll' (2012) *Metro* (Toronto), 12 April: 18

Franck, I. and Brownstone, D. (1992) *The Green Encyclopedia*, New York: Prentice Hall General Reference.

Frank, D.J. (1997) 'Science, nature and the globalization of the environment', 1870–1990, *Social Forces*, 76(2): 409–35.

Frank, D.J. (2002) 'The origins question: building global institutions to protect nature', in A.J. Hoffman and M.J. Ventresca (eds) *Organizations, Policy and the Natural Environment*, Stanford, CA: Stanford University Press.

Frank, D.J., Hironaka, A. and Schofer, E. (2000) 'The nation state and the natural environment over the twentieth century', *American Sociological Review*, 65(1): 96–116.

Freudenburg, W.R. (1997) 'Contamination, corrosion and the social order: an overview', *Current Sociology*, 45(3): 19–39.

Freudenburg, W.R. (2000) 'Social constructions and social constrictions: toward analyzing the social construction of "the naturalized" as well as "the natural"', in G. Spargaaren, A.P.G. Mol and F.H. Buttel (eds) *Environment and Global Modernity*, London: Sage.

Freudenburg, W.R. (2001) 'Risk, responsibility and recreancy', in G. Böhm, J. Nerb, T. McDaniels and H. Speda (eds) *Research in Social Problems and Public Policy, Volume 9* ('Environmental Risks: Perception, Evaluation and Management'): 87–108.

Freudenburg, W.R. and Alario, M. (2007) 'Weapons of mass destruction: magicianship, misdirection, and the dark side of legitimation', *Sociological Forum*, 22(2): 146–73.

Freudenburg, W.R. and Gramling, R. (1989) 'The emergence of environmental sociology: contributions of Riley E. Dunlap and William R. Catton, Jr.', *Sociological Inquiry*, 59(4): 439–52.

Freudenburg, W.R. and Gramling, R.B. (2011) *Catastrophe in the Making: The Engineering of Katrina and the Disasters of Tomorrow*, Washington, DC: Island Press.

Freudenburg, W.R. and Jones, T.R. (1991) 'Attitudes and stress in the presence of technological risk', *Social Forces*, 69(4): 1143–68.

Freudenburg, W.R. and Pastor, S. (1992) 'Public responses to technological risks: toward a sociological perspective', *The Sociological Quarterly*, 33(3): 389–412.

Freudenburg, W.R. and Wilkinson, R.C. (2008) 'Equity and the environment: a pressing need and a new step forward', in R.C. Wilkinson and W.R. Freudenburg (eds) *Equity and the Environment* (Volume 15, *Research in Social Problems and Public Policy*), Amsterdam: JAI Press/Elsevier.

Friedland, R. and Mohr, J. (2004) 'The cultural turn in American sociology', in R. Friedland and J. Mohr (eds) *Matters of Culture: Sociology in Practice*, New York: Cambridge University Press.

Friedman, S.M. (1983) 'Environmental reporting: a problem child of the media', *Environment*, 25(10): 24–9.

Friedman, S.M. (1984) 'Environmental reporting before and after TMI', *Environment*, 26(10): 4–5; 34.

Gamson, W.A. and Modigliani, A. (1989) 'Media discourse and public opinion on nuclear power', *American Journal of Sociology*, 95(1): 1–37.

Gamson, W.A. and Wolfsfeld, G. (1993) 'Movements and media as interacting systems', *Annals (AAPSS)*, 528: 114–25.

Gamson, W.A., Croteau, D., Haynes, W. and Sasson, T. (1992) 'Media images and social construction of reality', *Annual Review of Sociology*, 18: 373–93.

Gans, H.J. (1979) *Deciding What's News: A Study of CBS Evening News, NBC Nightly News, Newsweek and Time*, New York: Pantheon Books.

'Gas goes boom' (2013) *The Economist*, 2 June. Online. Available HTTP: http://www.economist.com/node/21556291 (accessed 30 October, 2013).

Gelcich, S., Edwards-Jones, G., Kaiser, M.J. and Watson, E. (2005) 'Using discourses for policy evaluation: the case of marine common property rights in Chile', *Society and Natural Resources*, 18(4): 377–91.

Gibbs, W.W. (2005) 'Obesity: an overblown epidemic?' *Scientific American*, 292(6): 70–7.

Giddens, A. (1981) *A Contemporary Critique of Historical Materialism*, Berkeley, CA: University of California Press.

Giddens, A. (2009) *The Politics of Climate Change*, Cambridge: Polity Press.

Giles, Erica (2012) 'Fracking fuels energy debate', *ScienceNews for Kids*, 18 July.

Gitlin, T. (1980) *The Whole World Is Watching: Mass Media in the Making or Unmaking of the New Left*, Berkeley and Los Angeles, CA: University of California Press.

Global Biodiversity Forum (1997) *Report of the Seventh Global Diversity Forum*, 6–8 June, Harare, Zimbabwe.

Goffman, E. (1974) *Frame Analysis: An Essay on the Organization of Experience*, Cambridge, MA: Harvard University Press.

Goldblatt, D. (1996) *Social Theory and the Environment*, Cambridge: Polity Press.

Goldman, M. and Schurman, R.A. (2000) 'Closing the "great divide": new social theory on society and nature', *Annual Review of Sociology*, 26(1): 563–84.

Goldsmith, E. and Allen, R.P. (with contributions from Michael Allaby, John Davoli and Sam Lawrence) (1972) 'A blueprint for survival', *The Ecologist*, 2(1).

Goodall, C. (2011) 'Peak stuff? Did the UK reach a maximum use of material resources in the early part of the last decade?' 13 October. Online. Available HTTP: http://www.carboncommentary.com/wp-content/uploads/2011/10/Peak_Stuff_17.10.11.pdf (accessed 29 August, 2013).

Goodell, J. (2012) 'The big fracking bubble: the scam behind the gas boom', *Rolling Stone*, 15 March.

Gooderham, M. (1994) 'Scientific group urges species inventory', *Globe & Mail*, 23 February: A12.

Goodspeed, P. (2005) 'Violent clashes threaten tsunami relief', *National Post* (Toronto), 19 July: A14.

Gore, A. (2009) *Our Choice: A Plan to Solve the Climate Crisis*, Emmaus, PA: Rodale.

Gottlieb, R. (1993) *Forcing the Spring: The Transformation of the American Environmental Movement*, Washington, DC: Island Press.

Gould, K., Pellow, D.N. and Schnaiberg, A. (2004) 'Interrogating the treadmill of production: everything you wanted to know about the treadmill but were afraid to ask', *Organization & Environment*, 17(3): 296–316.

Gould, K.A., Pellow, D.N. and Schnaiberg, A. (2008) *The Treadmill of Production: Injustice and Unsustainability in the Global Economy*, Boulder, CO: Paradigm Publishers.

Gould, K.A., Weinberg, A.S. and Schnaiberg, A. (1993) 'Legitimating impatience: Pyrrhic victories of the modern environmental movement', *Qualitative Sociology*, 16(3): 207–46.

Grant, M. (1916) *The Passing of the Great Race*, New York: Scribner.

Grayson, R. (2006) 'The first Earth Day'. *OTBKB (Only the Blog Knows Brooklyn)*, 22 April. Online. Available HTTP: http://onlytheblogknowsbrooklyn.com/2006/04/22/april_22_1970_t/ (accessed 29 August, 2013).

Greenpeace (1991) *Toxic Threat to Indian Lands*, Washington, DC: Greenpeace.

Greider, T. and Garkovitch, L. (1994) 'Landscapes: the social construction of nature and the environment', *Rural Sociology*, 59(1): 1–24.

Grumbine, R.E. (1992) *Ghost Bears: Exploring the Biodiversity Crisis*, Washington, DC: Island Press.

Grundmann, R. and Stehr, N. (2010) 'Climate change: what role for sociology?', *Current Sociology*, 58(6): 897–910.

Gusfield, J.R. (1984) 'On the side: practical action and social constructivism in social problems theory', in J.I. Kitsuse (ed.) *Studies in the Sociology of Social Problems*, Norwood, NJ: Ablex.

Gustafasson, K. (2011) 'Made in conflict: local residents' construction of a local environmental problem', *Local Environment*, 16(7): 655–70.

Haas, P. (1990) *Saving the Mediterranean: The Politics of International Environmental Cooperation*, New York: Columbia University Press.

Haas, P. (1992) 'Obtaining international protection through epistemic consensus', in I.H. Rowlands and M. Greene (eds) *Global Environmental Change and International Relations*, Basingstoke: Macmillan.

Hagen, J.B. (1992) *An Entangled Bank: The Origins of Ecosystem Ecology*, New Brunswick, NJ: Rutgers University Press.

Hajer, M.A. (1995) *The Politics of Environmental Discourse: Ecological Modernization and the Policy Process*, Oxford: Clarendon Press.

Hajer, M.A. and Versteeg, W. (2005) 'A decade of discourse analysis of environmental politics: achievements, challenges, perspectives', *Journal of Environmental Policy & Planning*, 7(3): 75–84.

Halper, Mark (2012) 'Converting danger into energy: Rwanda to divert lake's lethal gases to power plant,' *Smart Planet*, 14 February. Online. Available HTTP: http://www.smartplanet.com/blog/intelligent-energy/converting-danger-into-energy-rwanda-to-divert-lakes-lethal-gases-to-power-plant/13117 (accessed 2 September, 2013).

Hamilton, J. (2012a) 'Medical records could yield answers on fracking', National Public Radio ('Morning Edition'), 16 May. Online. Available HTTP: http://www.npr.org/2012/05/16/151762133/medical-records-could-yield-answers-on-fracking (accessed 2 September, 2013).

Hamilton, J. (2012b) 'Town's effort to link fracking and illness falls short', National Public Radio ('All Things Considered'), 16 May. Online. Available HTTP: http://www.npr.org/2012/05/16/152204584/towns-effort-to-link-fracking-and-illness-falls-short (accessed 2 September, 2013).

Hannigan, J.A. (1985) *Laboured Relations: Reporting Industrial Relations News in Canada*, Toronto: Centre for Industrial Relations, University of Toronto.

Hannigan, J.A. (1995) *Environmental Sociology: A Social Constructionist Perspective*, London and New York: Routledge.

Hannigan, J. (2002) 'Cultural analysis and environmental theory: an agenda', in R.E. Dunlap, F.H. Buttel, P. Dickens and A. Gijswijt (eds) *Sociological Theory and the Environment: Classical Foundations, Contemporary Insights*, Lanham, MD: Rowman & Littlefield.

Hannigan, J. (2005) 'The new urban political economy', in H.H. Hiller (ed.) *Urban Canada: Sociological Perspectives*, Toronto: Oxford University Press.

Hannigan, J. (2011) 'Social challenges: causes, explanations and solutions', in T. Fitzpatrick (ed.) *Understanding the Environment and Social Policy*, Bristol: The Policy Press.

Hannigan, J. (2012a) *Disasters Without Borders: The International Politics of Natural Disasters*, Cambridge: Polity Press.

Hannigan, J. (2012b) 'Grasping the nettle: sociology, political economy and carbon dependency: a response to Redclift', *International Review of Social Research*, 2(2): 73–80.

Hansen, A. (1991) 'The media and the social construction of the environment', *Media, Culture & Society*, 13(4): 443–58.

Hansen, A. (1993a) 'Introduction', in A. Hansen (ed.) *The Mass Media and Environmental Issues*, Leicester: Leicester University Press.

Hansen, A. (1993b) 'Greenpeace and press coverage of environmental issues', in A. Hansen (ed.) *The Mass Media and Environmental Issues*, Leicester: Leicester University Press.

Hansen, James (2008) 'Global warming twenty years later: tipping points near'. Online. Available HTTP: http://www.columbia.edu/~jeh1/2008/TwentyYearsLater_20080623.pdf (accessed 1 September, 2013).

Hardin, G. (1978) 'The tragedy of the commons', *Science*, 162: 1241–52.

Harper, C.L. (2008) *Environment and Society: Human Perspectives on Environmental Issues* (fourth edition), Upper Saddle River, NJ: Prentice Hall.

Harremoës, P., Gee, D., MacGarvin, M., Stirling, A., Keys, J., Wynne, B. and Vaz, S.G. (2002) 'Twelve late lessons', in P. Harremoës, D. Gee, M. MacGarvin, A. Stirling, J. Keys, B. Wynne and S.G. Vaz (eds), *The Precautionary Principle in the 20th Century: Late Lessons From Early Warnings*, London: Earthscan.

Harris, P.G. (2011) 'Climate change', in G. Kütting (ed.) *Global Environmental Politics: Concepts, Theories and Case Studies*, London and New York: Routledge.

Harrison, K. and Hoberg, G. (1991) 'Setting the environmental agendas in Canada and the United States: the case of dioxin and radon', *Canadian Journal of Political Science*, 24(1): 3–27.

Harrison, K. and Hoberg, G. (1994) *Risk, Science, and Politics: Regulating Toxic Substances in Canada and the United States*, Montreal and Kingston: McGill-Queen's University Press.

Hart, D.M. and Victor, D.G. (1993) 'Scientific elites and the making of U.S. policy for climate change research, 1957–1974', *Social Studies of Science*, 23(4): 643–80.

Harvey, D. (1974) 'Population, resources and the ideology of science', *Economic Geography*, 50(3): 265–77.

Harvey, D. (1989) *The Condition of Postmodernity*, Oxford: Blackwell.

Hawkins, A. (1993) 'Contested ground: international environmentalism and global climate change', in R.D. Lipchutz and K. Conca (eds) *The State and Social Power in Global Environmental Politics*, New York: Columbia University Press.

Hays, S. (1959) *Conservation and the Gospel of Efficiency: The Progressive Conservation Movement*, Cambridge, MA: Harvard University Press.

Heimer, C. (1998) 'Social structure, psychology and the estimation of risk', *Annual Review of Sociology*, 14(1): 491–519.

Hengeveld, R. (2012) *Wasted World: How Our Consumption Challenges the Planet*, Chicago: The University of Chicago Press.

Henninger, D. (2005) 'From spin city to fat city', *The Wall Street Journal*, 6 May: A14.

Herndl, C.G. and Brown, S.C. (1996) 'Introduction', in C.G. Herndl and S.C. Brown (eds) *Green Culture: Environmental Rhetoric in Contemporary America*, Madison, WI: University of Wisconsin Press.

Heyerdahl, T. (1958) *AKU-AKU: The Secret of Easter Island*, New York: Rand-McNally & Co.

Higgins, R.R. (1993) 'Race and environmental equity: an overview of the environmental justice issue in the policy process', *Polity*, 26(2): 281–300.

Higgins, V. and Natalier, K. (2004) 'Governing environmental harms in a risk society', in R. White (ed.) *Controversies in Environmental Sociology*, Cambridge: Cambridge University Press.

Higham, J. (1963) *Strangers in the Land: Patterns of American Nativism, 1860–1925*, New York: Atheneum.

Hilgartner, S. (1992) 'The social construction of risk objects: or how to pry open networks of risk', in J.F. Short Jr. and L. Clarke (eds) *Organizations, Uncertainties and Risk*, Boulder, CO: Westview Press.

Hilgartner, S. and Bosk, C.L. (1988) 'The rise and fall of social problems: a public arenas model', *American Journal of Sociology*, 94(1): 53–78.

Hill, D. (2013) 'Fracking opinions at tipping point', *The Hill*, 16 April. Online. Available HTTP: http://thehill.com/opinion/columnists/david-hill/295707-fracking-opinions-at-tipping-point (accessed 2 September, 2013).

Hites, R.A., Foran, J.A., Carpenter, D.O., Hamilton, M.C., Knuth, B.A. and Schwager, S.J. (2004) 'Global assessment of organic contaminants in farmed salmon', *Science*, 303, (5655), 9 January: 226–9.

Hochberg, L. (1980) 'Environmental reporting in boomtown Houston', *Columbia Journalism Review*, 19(2): 71–4.

Hoffmann, J. (2004) 'Social and environmental influences on endangered species: a cross-national study', *Sociological Perspectives*, 47(1): 79–107.

Hofrichter, R. (1993) 'Introduction', in R. Hofrichter (ed.) *Toxic Struggles: The Theory and Practice of Environmental Justice*, Philadelphia, PA: New Society Publishers.

Hofstadter, R. (1959) *Social Darwinism in American Thought*, New York: George Braziller.

Homeland Security News (2012) 'Fracking did not cause East Coast quakes, doubts linger', 13 January. Online. Available HTTP: http://www.homelandsecuritynewswire.com/dr20120113-fracking-did-not-cause-east-coast-quake-doubts-linger (accessed 3 February, 2014).

Horlick-Jones, T. (2004) 'Experts in risk? . . . do they exist?', *Health, Risk & Society*, 6(2): 107–14.

Howard, L.E. (1953) *Sir Albert Howard in India*, London: Faber & Faber.

Howarth, R.W., Ingraffea, A. and Engelder, T. (2011) 'Should fracking stop?', *Nature*, 477 (September): 271–5.

Howenstine, E. (1987) 'Environmental reporting: shift from 1970 to 1982', *Journalism Quarterly*, 64(4): 842–6.

Huber, J. (1982) *Die Verlorene Unschulde der Okologie: Neue Technologien und Superindustrielle Entwicklung* (The Lost Innocence of Ecology: New Technologies and Superindustrial Development), Frankfurt am Main: Fischer.

Huber, J. (1985) *Die Regenbogengesellschaft: Okologie und Sozialpolitik* (The Rainbow Society: Ecology and Social Policy), Frankfurt am Main: Fischer.

Hudson, W.H. (1944 [1916]) *Green Mansions: A Romance of the Tropical Forest*, New York: Random House.

Hulme, M. (2009) *Why We Disagree about Climate Change: Understanding Controversy, Inaction and Opportunity*, Cambridge: Cambridge University Press.

Humphrey, C.R., Lewis, T.L. and Buttel, F.H. (2003) 'Introduction: the development of environmental sociology', in C.R. Humphrey, T.L. Lewis and F.H. Buttel (eds) *Environment, Energy, and Society: Exemplary Works*, Belmont, CA: Wadsworth/Thomson Learning.

Hunt, S.A., Benford, R.D. and Snow, D.A. (1994) 'Identity fields: framing processes and the social construction of movement identities', in E. Larana, H. Johnston and J.R. Gusfield (eds) *New Social Movements: From Ideology to Identity*, Philadelphia, PA: Temple University Press.

Huntington, E. (1917) 'Climatic change and agricultural extension as elements in the fall of Rome', *Quarterly Journal of Economics*, 31(2): 173–208.

Hussain, Y. (2012) 'Gas industry tackles issues on fracking', *National Post* (Toronto), 23 March: FP6.

Ibarra, P.R. and Kitsuse, J.I. (1993) 'Vernacular constituents of moral discourse: an interactionist proposal for the study of social problems', in J. A. Holstein and G. Miller (eds) *Reconsidering Social Constructionism: Debates in Social Problems Theory*, New York: Aldine de Gruyter.

IPCC (2007) *Climate Change 2007: Synthesis Report: Contribution of Working Groups I, II and III to the Fourth Assessment Report of the Intergovernmental Panel on Climate Change*, Geneva, Switzerland.

Inkeles, A. and Smith, D.H. (1974) *Becoming Modern: Individual Change in Six Developing Countries*, Cambridge, MA: Harvard University Press.

IRIN (2007) 'DRC: Lake Kivu – a time bomb or source of energy?' *Humanitarian News and Analysis*, 15 August. Online. Available HTTP: http://www.irinnews.org/report/73738/drc-lake-kivu-a-time-bomb-or-source-of-energy (accessed 30 October, 2013).

Irwin, A. (2001) *Sociology and the Environment*, Cambridge: Polity Press.

Isaacs, C. (1994) 'Biodiversity may spur profits', *The Financial Post* (Toronto), 28 January: 17.

Jain, P. (2011) 'Club of Rome', in D. Mulvaney (general editor) and P. Robbins (series editor), *Green Politics: An A-to-Z Guide*, Thousand Oaks, CA: Sage.

Jamison, A. (1996) 'The shaping of the global environmental agenda: the role of non-governmental organizations', in S. Lash, B. Szerszynski and B. Wynne (eds) *Risk, Environment and Modernity: Towards a New Ecology*, London: Sage Publications.

Järvikoski, T. (1996) 'The relation of nature and society in Marx and Durkheim', *Acta Sociologica*, 39(1): 73–86.

Jasanoff, S. (1986) *Risk Management and Political Culture: A Comparative Study of Science in Policy Context*, New York: Russell Sage Foundation.

Jasanoff, S. (1990) *The Fifth Branch: Science Advisors as Policymakers*, Cambridge, MA and London: Harvard University Press.

Jenness, V. (1993) *The Prostitutes' Rights Movement in Perspective*, Hawthorne, NY: Aldine de Gruyter.

Johnson, I. (2013) 'China embarking on a vast program of urbanization', *New York Times*, 16 June: 1; 10.

'Joining together for Justice' (2004) *Sierra*, 89(3), May–June: 57.

'Jonathan Deal warns against fracking' (2013) *Recent News from Aberdeen* [blog], 31 March. Online. Available HTTP: http://aberdeencape.org/jonathan-deal-warns-against-fracking/ (accessed 2 September, 2013).

Jones, S. (2002) 'Social constructionism and the environment: through the quagmire', *Global Environmental Change* 12(4): 247–51.

Kalan, H. and Peck, B. (2005) 'South African perspectives on transnational environmental justice networks', in D.N. Pellow and R.J. Brulle (eds) *Power, Justice and the Environment: A Critical Appraisal of the Environmental Justice Movement*, Cambridge, MA: The MIT Press.

Kalman, J. (2013) 'Chinese struggle through "airpocalypse" smog', *The Observer*, 16 February. Online. Available HTTP: http://www.theguardian.com/world/2013/feb/16/chinese-struggle-through-airpocalypse-smog (accessed 3 February, 2014).

Kaminstein, D.S. (1988) 'Toxic talk', *Social Policy*, 19(2): 5–10.

Katz, E. and Lazarsfeld, P. (1955) *Personal Influence: The Part Played by People in the Flow of Mass Communication*, Glencoe, IL: Free Press.

Kebede, A. (2005) 'Grassroots environmental organizations in the United States: a Gramscian analysis', *Sociological Inquiry*, 75(1): 81–108.

Kellert, S.R. (1986) 'Social and perceptual factors in the preservation of animal species', in B.G. Norton (ed.) *The Preservation of Species: The Value of Biological Diversity*, Princeton, NJ: Princeton University Press.

Kellow, A. (2007) *Science and Public Policy: The Virtuous Corruption of Virtual Environmental Science*, Cheltenham: Edward Elgar.

Kidner, D.W. (2000) 'Fabricating nature: a critique of the social construction of nature', *Environmental Ethics*, 22(4): 339–57.

Killingsworth, M.J. and Palmer, J.S. (1992) *Ecospeak: Rhetoric and Environmental Politics in America*, Carbondale and Edwardsville, IL: Southern Illinois University Press.

Killingsworth, M.J. and Palmer, J.S. (1996) 'Millennial ecology: the apocalyptic narrative from *Silent Spring* to global warming', in C.G. Herndl and S.C. Brown (eds) *Green Culture: Environmental Rhetoric in Contemporary America*, Madison, WI: University of Wisconsin Press.

Kinchy, A.J. and Kleinman, D.L. (2003) 'Organizing credibility: discursive and organizational orthodoxy on the borders of ecology and politics', *Social Studies of Science*, 33(6): 869–96.

Kingdon, J.W. (1984) *Agendas, Alternatives and Public Policies*, Boston, MA: Little, Brown and Co.

Kitsuse, J.I., Murase, A.E. and Yamamura, Y. (1984) 'The emergence and institutionalization of an educated problem in Japan', in J. W. Schneider and J.I. Kitsuse (eds) *Studies in the Sociology of Social Problems*, Norwood, NJ: Ablex.

Kowalok, M. (1993) 'Research lessons from acid rain, ozone depletion and global warming', *Environment*, 35(60): 13–20; 35–8.

Krimsky, S. (1979) 'Regulating recombinant DNA research', in D. Nelkin (ed.) *Controversy: Politics of Technical Decisions*, Beverly Hills, CA: Sage.

Krimsky, S. and Plough, A. (1988) *Environmental Hazards: Communicating Risks as a Social Process*, Dover, MA: Auburn House.

Kroll-Smith, S., Couch, S.R. and Marshall, B.K. (1997) 'Sociology, extreme environments and social change', *Current Sociology*, 45(3): 1–18.

Kunst, M. and Witlox, N. (1993) 'Communication and the environment', *Communication Research Trends*, 13(4): 1–31.

Kwa, C. (1993) 'Radiation ecology, systems and the management of the environment', in M. Shortland (ed.) *Science and Nature: Essays in the History of the Environmental Sciences*, Oxford: British Society for the History of Science.

Lacey, C. and Longman, D. (1993) 'The press and public access to the environment and development debate', *The Sociological Review*, 41(2): 207–43.

Lahsen, M. (2013) 'Anatomy of dissent: a cultural analysis of climate skepticism', *American Behavioral Scientist*, 57(6): 732–53.

Lappé, F.M. and Collins, J., with Cary Fowler (1977) *Food First: Beyond the Myth of Scarcity*, Boston: Houghton Mifflin.

Larson, B.M.H., Nerlich, B. and Wallis, P. (2005) 'Metaphors and biorisks: the war on infectious diseases and invasive species', *Science Communication*, 26(3): 243–68.

Lash, S. and Wynne, B. (1992) 'Introduction' to U. Beck, *The Risk Society*, London: Sage.

Lash, S., Szersynski, B. and Wynne, B. (eds) (1996) *Risk, Environment and Modernity: Towards a New Ecology*, London: Sage.

Lawson, S. (1994) 'Farm widow refights old pipeline foe', *The Toronto Star*, 22 January: C-6.

Lear, L.K. (1993) 'Rachel Carson's *Silent Spring*', *Environmental History Review*, 17: 23–48.

Lee, C. (ed.) (1992) *Proceedings, First National People of Color Environmental Leadership Summit*, New York: United Church of Christ Commission for Racial Justice, December.

Lee, D.C. (1980) 'On the Marxian view of the relationship between man and nature', *Environmental Ethics*, 2(1): 3–16.

Lee, N. and Motzkau, J. (2013) 'Varieties of biosocial imagination: reframing responses to climate change and antibiotic resistance', *Science, Technology, & Human Values*, 38(4): 447–69.

Leopold, A. (1964 [1949]) *A Sand County Almanac*, New York: Oxford University Press.

Lerner, D. (1958) *The Passing of Traditional Society: Modernizing the Middle East*, Glencoe, IL: The Free Press.

Lever-Tracy, C. (2008a) 'Global warming and sociology', *Current Sociology*, 56(3): 445–66.

Lever-Tracy, C. (2008b) 'Global warming and sociology: reply', *Current Sociology*, 56(3): 485–91.

Liberatore, A. (1992) 'Facing global warming: the interactions between science and policy making in the European community', paper presented at the Symposium, 'Current Developments in Environmental Sociology', International Sociological Association Thematic Group on 'Environment and Society', Woudschoten, the Netherlands, June.

Lidskog, R. (1993) 'Review of U. Beck, *The Risk Society* [London: Sage, 1992]', *Acta Sociologica*, 36(4): 400–3.

Lidskog, R. (2001) 'The re-naturalization of society? Environmental challenges for sociology', *Current Sociology*, 49(1): 113–36.

Lim, L. (2013) 'Beijing's airpocalypse spurs pollution controls, public pressure', National Public Radio ('Morning Edition'), 14 January. Online. Available HTTP: http://www.npr.org/2013/01/14/169305324/beijings-air-quality-reaches-hazardous-levels (accessed 2 September, 2013).

Lindblom, C.E. and Cohen, D.H. (1979) *Usable Knowledge: Social Science and Social Problem Solving*, New Haven, CT: Yale University Press.

Lipschutz, R.D. (1996) *Global Civil Society and Global Environmental Governance: The Politics of Nature from Place to Planet*, Albany: SUNY Press.

Litfin, K. (1994) *Ozone Discourse*, New York: Columbia University Press.

'Little evidence to link fracking to earthquake risk, U.S. study says' (2012) *National Post*, 16 June: FP4.

Lockie, S. (1997) 'Chemical risk and the self-calculating farmer: diffuse chemical use in Australian broadacre farming systems', *Current Sociology*, 45(3): 81–97.

Lockie, S. (2004) 'Social nature: the environmental challenge to mainstream social theory', in R. White (ed.) *Controversies in Environmental Sociology*, Cambridge: Cambridge University Press.

Lomborg, B. (2001) *The Skeptical Environmentalist: Measuring the Real State of the World*, Cambridge: Cambridge University Press.

London, J. (2004 [1903]) *The Call of the Wild: Complete Text with Introduction, Historical Contexts, Critical Essays*, edited by Earl J. Wilcox, Boston: Houghton Mifflin Co.

Lorimer, J. (2006) 'What about the nematodes? Taxanomic partialities in the scope of UK biodiversity conservation', *Social and Cultural Geography*, 7(4): 539–58.

Louka, E. (2002) *Biodiversity and Human Rights: The International Rules for the Protection of Biodiversity*, Ardsley, NY: Transnational Publishers.

Lovejoy, T.E. (1986) 'Species leave the ark: one by one', in B.G. Norton (ed.) *The Preservation of Species: The Value of Biological Diversity*, Princeton, NJ: Princeton University Press.

Lovell, J. (2005) 'Kilimanjaro's global warming: Africa's tallest mountain stripped of snow a wake-up call to G8, environmentalist says', *Globe & Mail*, 15 March: A1, A16.

Lovelock, J. (1979) *Gaia: A New Look at Life on Earth*. Oxford: Oxford University Press.

Lowe, P.D. and Goyder, J. (1983) *Environmental Groups in Politics*, London: Allen & Unwin.

Lowe, P.D. and Morrison, D. (1984) 'Bad news or good news: environmental politics and the mass media', *The Sociological Review*, 32(1): 75–90.

Lupton, D. (2004) 'A grim health future: food risks in the Sydney press', *Health, Risk and Society*, 6(2): 187–200.

Lutts, R.H. (1990) *The Nature Fakers: Wildlife, Science and Sentiment*, Golden, CO: Fulcrum.

Lynch, B.D. (1993) 'The garden and the sea: US Latino environmental discourses and mainstream environmentalism', *Social Problems*, 40(1): 108–24

McComas, K. and Shanahan, J. (1999) 'Telling stories about global climate change', *Communication Research*, 26(1): 30–57.

McCright, A.M. and Dunlap, R.E. (2003) 'Defeating Kyoto: the conservative movement's impact on US climate change policy', *Social Problems*, 50(3): 348–73.

Macdonald, D. (1991) *The Politics of Pollution: Why Canadians are Failing their Environment*, Toronto: McClelland & Stewart.

McDonald, K. (2004) 'Oneself as another: from social movement to experience movement', *Current Sociology*, 52(4): 575–93.

MacDonald, K.I. (2010) 'The devil is in the (bio)diversity: private sector "engagement" and the restructuring of biodiversity conservation', *Antipode* 42(3): 513–50.

Macfarlane, R. (2005) '4x4s are killing my planet', *The Guardian*, 4 June: 6–7.

McIntosh, R.P. (1985) *The Background of Ecology: Concept and Theory*, Cambridge: Cambridge University Press.

McIntyre, S. (2005) 'Revisiting the stick', *National Post* (Toronto), 17 June: FP-19.

MacKenzie, D. (2012) 'The Limits of Growth revisited,' *New Scientist*, 213(2846): 38–41.

McKibben, B. (2006) *The End of Nature*, New York: Random House.

McManus, J. (ed.) (1989) 'Planet of the year', *Time*, 2 January.

McNeely, J.A. (1992) 'Review of R. Tobin, *The Expendable Future: U.S Politics and the Protection of Biological Diversity* [Durham, NC: Duke University Press, 1991]', *Environment*, 34(2): 25–7.

McNeely, J.A., Miller, K.R., Reid, W.V., Mittermeier, R.A. and Werner, T.B. (1990a) *Conserving the World's Biological Diversity*, Gland, Switzerland and Washington, DC: World Conservation Union, World Resources Institute, Conservation International, World Wildlife Fund, and World Bank.

McNeely, J.A., Miller, K.R., Reid, W.V., Mittermeier, R.A. and Werner, T.B. (1990b) 'Strategies for conserving biodiversity', *Environment*, 32(3): 16–20; 36–40.

Mahon, T. (1985) *Charged Bodies: People, Power and Paradox in Silicon Valley*, New York: New American Library.

Manes, C. (1990) *Green Rage: Radical Environmentalism and the Unmaking of Civilization*, Boston, MA: Little, Brown & Co.

Mann, C.C. and Plummer, M.L. (1992) 'The butterfly problem', *The Atlantic Monthly*, 229(1): 47–70.

Mannion, A. (1993) 'Review of W. Reid *et al.* (1993)', *Environmental Conservation*, 20(3): 286–7.

Marshall, M. (2011) 'How fracking caused earthquakes in the UK', *New Scientist*, 2 November. Online. Available HTTP: http://www.newscientist.com/article/dn21120-how-fracking-caused-earthquakes-in-the-uk.html#.Uu8AZXkSHwI (accessed 3 February, 2014).

Martel, N. (2005) 'Eco-lessons taught in a surfer-girl patois', *The New York Times*, 28 March: B1–2.

Martell, L. (1994) *Ecology and Society: An Introduction*, Cambridge: Polity Press.

Martens, S. (2007) 'Public participation with Chinese characteristics: citizen consumers in China's environmental management', in N.T. Carter and A.P.J. Mol (eds) *Environmental Governance in China*, London and New York: Routledge.

Masny, L. (2000) 'Ice cover melting worldwide', in L. Brown, M. Renner and B. Halwell (eds), *Vital Signs 2000*, New York and London: W.W. Norton & Company.

Mayer, E.L. (1992) 'Environmental racism', *Audubon*, January/February: 30–2.

Mazur, A. (1985) 'The mass media in environmental controversies', in A. Brannigan and S. Goldenberg (eds) *Social Responses to Technological Change*, Westport, CT: Greenwood Press.

Mazur, A. and Lee, J. (1993) 'Sounding the global alarm: environmental issues in the U.S. national news', *Social Studies of Science*, 23(4): 671–720.

Meadows, D.H., Meadows D.L., Randers, J. and W.W. Behrens III (1972) *The Limits to Growth*, Washington, DC: Universe Books.

Meadows, D.H., Randers, J. and Meadows, D.L. (2004) *Limits to Growth: The 30-Year Update*, White River Junction, VT: Chelsea Green Publishing.

Meadows, D.L. (1973) 'Introduction to the project', in D.L. Meadows and D.H. Meadows (eds) *Toward Global Equilibrium: Collected Papers*, Cambridge, MA: Wright-Allen Press, Inc.

Merchant, C. (1987) 'The theoretical structure of ecological revolutions', *Environmental Review*, 11(4): 265–74.

Mertig, A.G. and Dunlap, R.E. (2003) 'Environmentalism', in N.J. Smelser and P.B. Baltes (eds) *International Encyclopedia of the Social and Behavioral Sciences*, Kiddington: Elsevier Science.

Merton, R.H. and Nisbet, R.A. (1971) *Contemporary Social Problems (3rd edition)*, New York: Harcourt, Brace World.

Meyer, D.S. and Rohlinger, D.A. (2012) 'Big books and social movements: a myth of ideas and social change', *Social Problems*, 59(1): 136–53.

Milbrath, L.W. (1989) *Envisioning a Sustainable Society: Learning Our Way Out*, Albany: SUNY Press.

Millar, C.I. and Ford, L.D. (1988) 'Managing for nature: from genes to ecosystems', *BioScience*, 38(7): 456–7.

Miller, C.A. and Edwards, P.N. (2001) 'Introduction: the globalization of climate science and climate politics', in C.A. Miller and P.N. Edwards (eds) *Changing the Atmosphere: Expert Knowledge and Environmental Governance*, Cambridge, MA and London: The MIT Press.

Miller, V.D. (1993) 'Building on our past, planning our future: communities of color and the quest for environmental justice', in R. Hofrichter (ed.) *Toxic Struggles: The Theory and Practice of Environmental Justice*, Philadelphia, PA: New Society Publishers.

Mills, C.W. (1959) *The Sociological Imagination*, New York: Oxford University Press.

Milne, A. (1993) 'The perils of green pessimism', *New Scientist*, 138(1877), 12 June: 34–7.

Milton, K. (1991) 'Interpreting environmental policy: a social scientific approach', *Journal of Law and Society*, 18(1): 4–17.

Mitchell, R.B. (2010) *International Politics and the Environment*, Thousand Oaks, CA: Sage.

Modavi, N. (1991) 'Environmentalism, state and economy in the United States', *Research in Social Movements, Conflicts and Change*, 13: 261–73.

Mohai, P. (2008) 'Equity and the environmental justice debate', in R.C. Wilkinson and W.R. Freudenburg (eds) *Equity and the Environment* (Volume 15, *Research in Social Problems and Public Policy*), Amsterdam: JAI Press/Elsevier.

Mol, A.P.J. (1997) 'Ecological modernization: industrial transformation and environmental reform', in M. Redclift and G. Woodgate (eds) *The International Handbook of Environmental Sociology*, London: Edward Elgar.

Mol, A.P.J. and Carter, N.T. (2007) 'China's environmental governance in transition', in N.T. Carter and A.P.J. Mol (eds) *Environmental Governance in China*, London and New York: Routledge.

Mol, A.P.J and Spaargaren, G. (2000) 'Ecological modernisation theory in debate: a review', *Environmental Politics* 9(1): 17–49.

Mol, A.P.J. and Spaargaren, G. (2003) 'Towards a sociology of environmental flows: a new agenda for 21st century environmental sociology', paper presented at the International Conference on 'Governing Environmental Flows', Wageningen, the Netherlands, June 13–14.

Molotch, H. (1970) 'Oil in Santa Barbara and power in America', *Sociological Inquiry*, 40(1): 131–44.

Molotch, H. and Lester, M. (1975) 'Accidental news: the great oil spill as local occurrence and national event', *American Journal of Sociology*, 81(2): 235–60.

Monbiot, G. (2011) 'Peak stuff?' 3 November. Online. Available HTTP: http://www.monbiot.com/2011/11/03/peak-stuff/ (accessed 2 September, 2013).

Monteiro, G. (2005) 'Review essay; discourses on abortion, discourses on politics: two studies in the politics of discourse,' *Current Sociology*, 53(1): 157–68.

Mooney, C. (2011) 'The truth about fracking', *Scientific American*, 305 (November): 80–5.

Mooney, H.A. (2003) 'The ecology-policy interface', *Frontiers in Ecology and the Environment*, 1(1): 49.

Mooney, P.R. (1979) *Seeds of the Earth: A Private or Public Resource?* Ottawa: inter Pares for the Canadian Council for International Co-operation and the International Coalition for Development Action.

Morrison, D.E., Hornback, H.E. and Warner, W.H. (1972) 'The environmental movement: some preliminary observations and predictions', in W.R. Burch Jr., N.H. Cheek Jr. and L. Taylor (eds.) *Social Behavior, Natural Resources and the Environment*, New York: Harper and Row.

Murphy, P. and Maynard, M. (2000) 'Framing the genetic testing issue: discourse and cultural clashes among policy communities', *Science Communication*, 22(2): 133–53.

Murphy, R. (1994) *Rationality and Nature: A Sociological Inquiry into a Changing Relationship*, Boulder, CO: Westview Press.

Murphy, R. (2011) 'The challenge of anthropogenic climate change for the social sciences', *International Review of Social Research*, 1(3): 167–81.

Myerson, G. and Rydin, Y. (1996) *The Language of Environment: A New Rhetoric*, London: UCL Press.

Nadeau, R.L. (2006) *The Environmental Endgame: Mainstream Economics, Ecological Disaster, and Human Survival*, New Brunswick, NJ: Rutgers University Press.

Nash, R. (1967) *Wilderness and the American Mind*, New Haven, CT: Yale University Press.

Nash, R. (1977) 'The value of wilderness', *Environmental Review*, 1: 14–25.

Nash, R. (1989) *The Rights of Nature: A History of Environmental Ethics*, Madison, WI: University of Wisconsin Press.

Nelissen, N., Van Der Straaten, J. and Klinkers, L. (eds) (1997) *Classics in Environmental Studies: An Overview of Classic Texts in Environmental Studies*, Utrecht, the Netherlands: International Books.

Nelkin, D. (1987) 'The culture of science journalism', *Society*, 24(6): 17–25.

Nelkin, D. (1989) 'Communicating technological risk: the social construction of risk perception', *Annual Review of Public Health*, 10: 95–113.

Nicola, S. (2013) '500-year-old German law protects "beer purity"', *National Post*, 24 May: FP1, 6.

Norton, B.G. (1986) 'Introduction to Part 1', in B.G. Norton (ed.) *The Preservation of Species: The Value of Biological Diversity*, Princeton, NJ: Princeton University Press.

Noss, R. (1990) 'From endangered species to biodiversity', in K. Kohm (ed.) *Balancing on the Brink of Extinction*, Washington, DC: Island Press.

Novek, J. and Kampen, K. (1992) 'Sustainable or unsustainable development? An analysis of an environmental controversy', *Canadian Journal of Sociology*, 17(3): 249–73.

O'Riordan, T. (1976) *Environmentalism*, London: Pion.

Oliver, J.E. (2005) *Obesity: The Making of an American Epidemic*, Oxford: Oxford University Press.

Oreskes, N. (2004) 'Beyond the Ivory Tower: the scientific consensus on climate change', *Science*, 306(5702): 1686.

Oreskes, N. and Conway, E.M. (2010) *Merchants of Doubt*, New York: Bloomsbury.

Palmlund, I. (1992) 'Social drama and risk evaluation', in S. Krimsky and D. Golding (eds) *Social Theories of Risk*, Westport, CT: Praeger.

Papadakis, E. (1984) *The Green Movement in West Germany*, London, Canberra and New York: Croom Helm/St. Martin's Press.

Park, R.E. (1952) 'Human ecology', in R.E. Park (ed.) *Human Communities: The City and Human Ecology*, New York: The Free Press. (Originally published in 1936 in *American Journal of Sociology*, 42: 1–15.)

Parlour, J.W. and Schatzow, S. (1978) 'The mass media and public concern for environmental problems in Canada: 1960–1972', *International Journal of Environmental Studies*, 13(1): 9–17.

Parsons, H.L. (1977) *Marx and Engels on Ecology*, Westport, CT: Greenwood Press.

Paterson, M. (1994) 'The politics of climate change after UNCED', in C. Thomas (ed.), *Rio: Unravelling the Consequences*, Ilford: Frank Cass.

Paterson, M. (2000) 'Swampy fever: media constructions and direct action politics', in B. Seel, M. Paterson and B. Doherty (eds) *Direct Action: British Environmentalism*, London and New York: Routledge.

Payne, B.A. and Cluett, C. (2002) 'Environmental sociology', in R.E. Dunlap and W. Michelson (eds) *Handbook of Environmental Sociology*, Westport, CT: Greenwood Press.

Pearce, F. (1991) *Green Warriors: The People and Politics Behind the Environmental Revolution*, London: The Bodley Head.

Pearce, F. (2010) *The Climate Files: The Battle for Truth about Global Warming*, London: Guardian Books.

Pellow, D.N. (2000) 'Environmental inequality formation: toward a theory of environmental injustice', *American Behavioral Scientist*, 43(4): 581–601.

Pellow, D.N. (2007*) Resisting Global Toxics: Transnational Movements for Environmental Justice*, Cambridge, MA: The MIT Press.

Peluso, N.L. (1992) *Rich Forests, Poor People: Resource Control and Resistance in Java*, Berkeley, CA: University of California Press.

Peluso, N.L. (2003) 'Territorializing local struggles for resource control', in P. Greenough and A.L. Tsing (eds) *Nature in the Global South: Environmental Projects in South and Southeast Asia*, Durham, NC and London: Duke University Press.

Perrow, C. (1999) *Normal Accidents: Living with High Risk Technologies*, Princeton, NJ: Princeton University Press.

Perry, M. (1994) 'The toilet bowl they call Sydney Harbor', *Toronto Star*, 24 July: B-7.

Pettenger, M. (ed.) (2013) *The Social Construction of Climate Change*, Aldershot: Ashgate.

Peuhkuri, T. (2002) 'Knowledge and interpretation in environmental conflict: fish farming and eutrophication in the Archipelago Sea, SW Finland', *Landscape and Urban Planning*, 61(2): 157–68.

Picou, J.S. and Gill, D.A. (2000) 'The Exxon Valdez disaster as localized environmental catastrophe: dissimilarities to risk society theory', in M.J. Cohen (ed.) *Risk in the Modern Age: Social Theory, Science and Environmental Decision-Making*, Houndsmills, Basingstoke: Macmillan Press, Ltd. and New York: St. Martin's Press, Inc.

Pielke, R.A., Jr. (2004a) 'When scientists politicize science: making sense of controversy over *The Skeptical Environmentalist*', *Environmental Science & Policy*, 7: 405–17.

Pielke, R.A., Jr. (2004b) *The Honest Broker*, Cambridge: Cambridge University Press.

Plein, L.C. (1991) 'Popularizing biotechnology: the influence of issue definition', *Science, Technology & Human Values*, 16(4): 474–90.

'Pollution rife in Rio+20 host city' (2012) *Metro (Toronto)*, 21 June: 10.

Porritt, J. and Winner, D. (1988) *The Coming of the Greens*, London: Fontana.

Pretty, J., Ball, A.S., Benton, T., Guivant, J.S., Lee, D.R., Orr, D., Pfeffer, M.J. and Ward, H. (2007) 'Introduction to environment and society', in J. Pretty, A.S. Ball, T. Benton, J.S. Guivant, D.R. Lee, D. Orr, M. Pfeffer and H. Ward (eds), *The SAGE Handbook of Environment and Society*, London: SAGE Publications.

Raffaelli, D. (2004) 'How extinction patterns affect ecosystems', *Science*, 306(5699): 1141–2.

Rafter, N. (1992) 'Claims-making and socio-cultural context in the first US eugenics campaign', *Social Problems*, 35: 17–34.

Rajão, R. and Vurdubakis, T. (2013) 'On the pragmatics of inscription: detecting deforestation in the Brazilian Amazon', *Theory, Culture & Society*, 30(4): 151–77.

Randers, J. (2012) *2052: A Global Forecast for the Next Forty Years*, White River Junction, VT: Chelsea Green Publishing.

Randers, J. and Meadows, D.H. (1973) 'The carrying capacity of our global environment: a look at the ethical alternatives', in D.L. Meadows and D.H. Meadows (eds) *Toward Global Equilibrium: Collected Papers*, Cambridge, MA: Wright-Allen Press, Inc.

Redclift, M. (1984) *Development and the Environmental Crisis: Red or Green Alternatives?* New York: Methuen.

Redclift, M. (1986) 'Redefining the environmental "crisis" in the South', in J. Weston (ed.) *Red and Green: The New Politics of the Environment*, London: Pluto Press.

Redclift, M. (2000) 'Environmental social theory for a globalizing world economy', in G. Spaargaren, A.P.J. Mol and F.H. Buttel (eds) *Environment and Global Modernity*, London: Sage.

Redclift, M. (2009) 'The environment and carbon dependence: landscapes of sustainability and materiality', *Current Sociology*, 57(3): 369–88.

Redclift, M. (2011) 'The response of the hermeneutic social sciences to a "post carbon world"', *International Review of Social Research* 3: 155–66.

Redclift, M. and Woodgate, G. (1994) 'Sociology and the environment: discordant discourse?', in M. Redclift and T. Benton (eds) *Social Theory and the Global Environment*, London and New York: Routledge.

Redclift, M. and Woodgate, G. (eds) (1997) *The International Handbook of Environmental Sociology*, Cheltenham: Edward Elgar.

Regalado, A. (2005) 'Global warring: in climate debate, the "hockey stick" leads to a face-off', *The Wall Street Journal*, 14 February: A1; A13.

Reguly, E. (2013) 'Shale gives Obama climate leeway', *Globe & Mail*, 29 June: B1; B2.

Reid, W.V., McNeely, J.A., Turnstull, D.B., Bryant, D.A. and Winagrad, M. (1993) *Biodiversity Indicators of Policy Makers*. Washington, DC: World Resources Institute, October. Online. Available HTTP: http://pdf.wri.org/biodiversity_indicators_front.pdf (accessed 30 October, 2013).

Renn, O. (1992) 'Concepts of risk: a classification', in S. Krimsky and D. Golding (eds) *Social Theories of Risk*, Westport, CT: Praeger.

Reusswig, F. (2011) 'Sociological tasks in view of the transition to post-carbon societies. Also a comment to Michael Redclift', *International Review of Social Research*, 1(3): 189–95.

Rice, J. (2013) 'Further beyond the Durkheimian problematic: environmental sociology and the co-construction of the social and natural', *Sociological Forum*, 28(2): 236–60.

Rice, X. (2010) 'Rwanda harnesses volcanic gases from depths of Lake Kiva', *The Guardian*, 16 August. Online. Available HTTP: http:/www.theguardian.com/environment/2010/aug/16/rwanda-gas-lake-kivu

Richardson, M., Sherman, J. and Gismondi, M. (1993) *Winning Back the Words: Confronting Experts in an Environmental Public Hearing*, Toronto: Garamond Press.

Rifkin, J. (1992) *Beyond Beef: The Rise and Fall of Cattle Culture*, New York: Dutton.

Rittenhouse, C.A. (1991) 'The emergence of premenstrual syndrome as a social problem', *Social Problems*, 38(3): 412–25.

Rochon, T.R. (1998) *Culture Moves: Ideas, Activism and Changing Values*, Princeton, NJ: Princeton University Press.

Ronderos, M.T. (2004) 'Going it alone', *The Ecologist*, 34(2): 58–9.

Roth, C.E. (1978) 'Off the merry-go-round, on the escalator' in W.B. Stapp (ed.) *From Ought to Action in Environmental Education*, Columbus, OH: ERIE/SMEAC.

Roué, M. (2003) 'US environmental NGOs and the Cree. An unnatural alliance for the preservation of nature?', *International Social Science Journal*, 178 (December): 619–27.

Rubin, C.T. (1994) *The Green Crusade: Rethinking the Roots of Environmentalism*, New York: The Free Press.

Rudel, T.K., Roberts, J.T. and Carmin, J. (2011) 'Political economy of the environment', *Annual Review of Sociology*, 37: 221–38.

Rudiak-Gould, Peter (2012) 'Progress, decline and the public uptake of climate science', *Public Understanding of Science*, 5 June. Online. Available HTTP: http://pus.sagepub.com/content/early/2012/06/03/0963662512444682 (accessed 2 September, 2013).

Rugh, Peter (2013) 'Forest Service seeks to silence Smokey the Bear over fracking', *Waging Nonviolence*, 7 May. Online. Available HTTP: http://wagingnonviolence.org/feature/forest-service-seeks-to-silence-smokey-the-bear-over-fracking/ (accessed 2 September, 2013).

Ryan, C. (1991) *Prime Time Activism*, Boston: South End Press.

Rycroft, R.W. (1991) 'Environmentalism and science: politics and the pursuit of knowledge', *Knowledge: Creation, Diffusion, Utilization*, 13(2): 150–69.

Sachs, W. (1997) 'Sustainable development', in M. Redclift and G. Woodgate (eds) *The International Handbook of Environmental Sociology*, Cheltenham: Edward Elgar Publishing Ltd.

Sagan, C. (1991) 'Croesus and Cassandra', in A.B. Wolbarst (ed.) *Environment in Peril*, Washington and London: Smithsonian Institution Press.

Salter, L. (with the assistance of E. Levy and W. Leiss) (1988) *Science and Scientists in the Making of Standards*, Dordrecht: Kluwer Academic.

'Samsonite pulls luggage from Hong Kong stores' (2012) *Metro* (Toronto), 19 June: 24.

Sandman, P.M., Sachman, D.B., Greenberg, M. and Gochfeld, M. (1987) *Environmental Risk and the Press: An Exploratory Assessment*, New Brunswick, NJ: Transaction Books.

Schama, S. (1996) *Landscape and Memory*, Toronto: Vintage Canada.

Schlesinger, P. (1978) *Putting 'Reality' Together: BBC News*, London: Constable.

Schmitt, P.J. (1990) *Back to Nature: The Arcadian Myth in Urban America*, Baltimore, MA: Johns Hopkins University Press.

Schnaiberg, A. (1980) *The Environment: From Surplus to Scarcity*, New York: Oxford University Press.

Schnaiberg, A. (1993) 'Introduction: inequality once more, with (some) feeling', *Qualitative Sociology*, 16(3): 203–6.

Schnaiberg, A. (2002) 'Reflections on my 25 years before the mast of the environment and technology section', *Organization & Environment*, 15(1): 30–41.

Schnaiberg, A. (2003) 'American environmental studies', in N.J. Smelser and P.B. Baltes (eds) *International Encyclopedia of the Social and Behavioral Sciences*, Kiddington: Elsevier Science.

Schnaiberg, A. and Gould, K.A. (1994) *Environment and Society*, New York: St. Martin's Press.

Schoenfeld, A.C. (1980) 'Newspapers and the environment today', *Journalism Quarterly*, 57: 456–62.

Schoenfeld, A.C., Meier, R.F. and Griffin, R.J. (1979) 'Constructing a social problem: the press and the environment', *Social Problems*, 27(1): 38–61.

Schrepfer, S.R. (1983) *The Fight to Save the Redwoods: A History of Environmental Reform 1917–1978*, Madison, WI: The University of Wisconsin Press.

Schumacher, E.F. (1974) *Small is Beautiful*, London: Abacus.

Scotland, R. (1994) 'Marketing model helps rate brand performance', *The Financial Post*, (Toronto), 1 December: 23.

Scott, J. (2001) *Power*, Cambridge: Polity/Blackwell.

Sealey-Huggins, L., Rohloff, A. and Hulme, M. (2012) 'Book review symposium: John Urry, *Climate Change and Society* (2011)', *Sociology*, 46(3): 549–54.

Sears, P.B. (1964) 'Ecology as a subversive subject', *Bioscience*, 14(7): 11–13.

Seippel, O. (2002) 'Modernity, politics and the environment: a theoretical perspective', in R.E. Dunlap, F.H. Buttel and A. Gijswijt (eds) *Sociological Theory and the Environment: Classical Foundations, Contemporary Insights*, Lanham MD: Rowman & Littlefield.

Selznick, P. (1960) *The Organizational Weapon*, Glencoe, IL: Free Press.

Shackley, S. (2001) 'Epistemic lifestyles in climate change modeling', in C.A. Miller and P.N. Edwards (eds) *Changing the Atmosphere: Expert Knowledge and Environmental Governance*, Cambridge, MA and London: The MIT Press.

'Shale and gas resources are globally abundant' (2013) *Global Energy World*, 11 June. Online. Available HTTP: http://www.globalenergyworld.com/news/8253/Shale_oil_and_shale_gas_resources_are_globally_abundant.htm (accessed 2 September, 2013).

Shapiro, S. (1993) 'Rejoining the battle against noise pollution', *Issues in Science and Technology*, 9(3): 73–9.

Sheail, J. (1976) *Nature in Trust: The History of Nature Conservation in Britain*, Glasgow: Blackie.

Shenon, P. (1994) 'A Vietnamese goat is imperiled by fame', *The New York Times*, 29 November: A-6.

Shepard, P. (1969) 'Ecology and Man: a viewpoint', in P. Shepard and D. McKinly (eds) *The Subversive Science: Essays Toward an Ecology of Man*, Boston: Houghton Mifflin.

Shermer, Michael (2000) 'The evolution wars.' Online. Available HTTP: http://www.michaelshermer.com/2000/07/the-evolution-wars/ (accessed 2 September, 2013).

Shiva, V. (1990) 'Biodiversity, biotechnology and profit: the need for a People's Plan to protect biological diversity', *The Ecologist*, 20(2): 44–7.

Shiva, V. (1993a) 'Farmers' rights, biodiversity and international treaties', *Economic and Political Weekly* (Bombay), 28(4), 3 April: 555–60.

Shiva, V. and Holla-Bhar, R. (1993b) 'Intellectual piracy and the neem tree', *The Ecologist*, 23(6), November/December: 223–7.

Shogren, E. (2012) 'Fracking's methane trail: a detective story', National Public Radio ('Morning Edition'), 17 May. Online. Available HTTP: http://www.npr.org/2012/05/17/151545578/frackings-methane-trail-a-detective-story (accessed 2 September, 2013).

Simonis, U. (1989) 'Ecological modernization of industrial society: the strategic elements', *International Social Science Journal*, 121: 347–61.

Sinclair, U. (1906) *The Jungle*, New York: Doubleday, Page & Company.

Skodvin, T. (2000a) 'The ozone regime', in S. Andresen, T. Skodvin, A. Underdal and J. Wettestad (eds) *Science and Politics in International Environmental Regimes*, Manchester and New York: Manchester University Press.

Skodvin, T. (2000b) 'The International Panel on Climate Change', in S. Andresen, T. Skodvin, A. Underdal and J. Wettestad (eds) *Science and Politics in International Environmental Regimes*, Manchester and New York: Manchester University Press.

Slater, C. (2002) *Entangled Edens: Visions of the Amazon*, Berkeley, CA: University of California Press.

Smith, A. and Kern, F. (2007) 'The transitions discourse in the ecological modernisation of the Netherlands', SPRU (Science and Technology Policy Research), University of Sussex. Paper for the Earth Systems Governance Conference in Amsterdam, 24–26 May. Online. Available HTTP: http://www.2007amsterdamconference.org/Downloads/AC2007_SmithKern.pdf (accessed 9 July, 2012).

Smith, C. (1992) *Media and Apocalypse: News Coverage of the Yellowstone Forest Fires, Exxon Valdez Oil Spill and Loma Prieta Earthquake*, Westport, CT: Greenwood Press.

Smith, D. (2008) 'Editorial: beyond greed, fear and anger,' *Current Sociology*, 56(3): 347–50.

Smith, M. (1999) 'To speak of trees: social constructivism, environmental values, and the future of deep ecology', *Environmental Ethics*, 21(4): 359–76.

Smithson, M. (1989) *Ignorance and Uncertainty: Emerging Paradigms*, New York: Springer-Verlag.

Snow, C.P. (2013 [1961]) *Science and Government*. Cambridge, MS: Harvard University Press, 1961.

Snow, D.A., Rocheford, E.B. Jr., Warden, S.H. and Benford, R.D. (1986) 'Frame alignment processes, micromobilization and movement participation', *American Sociological Review*, 51: 464–81.

Solesbury, W. (1976) 'The environmental agenda: an illustration of how situations may become political issues and issues may demand responses from government: or how they may not', *Public Administration*, 54(4): 379–97.

Sontag, S. (1991) *Aids and its Metaphors*, Harmondsworth: Penguin.

Sorokin, P.A. (1964) [1928] *Contemporary Sociological Theories: Through the First Quarter of the Twentieth Century*, New York: Harper & Row.

Soulé, M. and Kohm, K. A. (1989) *Research Priorities for Conservation Biology*, Washington, DC: Island Press.

Soulé, M. and Lease, G. (1995) (eds) *Reinventing Nature? Responses to Postmodern Deconstruction*, Washington, DC: Island Press.

Spaargaren, G. (2000) 'Ecological modernization theory and the changing discourse on environment and modernity', in G. Spaargaren, A.P.J. Moland and F.H. Buttel (eds) *Environment and Global Modernity*, London: Sage.

Spaargaren, G. and Mol, A.P.J. (1992) 'Sociology, environment and modernity: ecological modernization as a theory of social change', *Society and Natural Resources*, 5: 323–44.

Spaargaren, G., Mol, A.P.J. and Buttel, F.H. (2000a) 'Introduction: globalization, modernity and the environment', in G. Spaargaren, A.P.J. Mol and F.H. Buttel (eds) *Environment and Global Modernity*, London: Sage.

Spaargaren, G., Mol, A.P.J. and Buttel, F.H. (eds) (2000b) *Environment and Global Modernity*, London: Sage.

Spector, M. and Kitsuse, J.I. (1973) 'Social problems: a reformation', *Social Problems*, 20: 145–59.

Spector, M. and Kitsuse, J.I. (1977) *Constructing Social Problems*, Menlo Park, CA: Cummings.

Spencer, J.W. and Triche, E. (1994) 'Media constructions of risk and safety: differential framings of hazard events', *Sociological Inquiry*, 64(2): 199–213.

Spray, S.L. and McGlothlin, K.L. (2003a) 'Introduction', in S.L. Spray and K.L. McGlothlin (eds) *Loss of Biodiversity*, Lanham, MD: Rowman & Littlefield.

Spray, S.L. and McGlothlin, K.L. (2003b) 'The global challenge: concluding thoughts on the loss of biodiversity', in S.L. Spray and K.L. McGlothlin (eds) *Loss of Biodiversity*, Lanham, MD: Rowman & Littlefield.

Staggenborg, S. (1993) 'Critical events and the mobilization of the pro-choice movement', *Research in Political Sociology*, 6: 319–45.

Stallings, R. (1990) 'Media discourse and the social construction of risk', *Social Problems*, 37: 80–95.

Stehr, N. and von Storch, H. (1995) 'The social construct of climate and climate change', *Climate Research*, 5: 99–105.

Stehr, N. and von Storch, H. (2010) *Climate and Society*, Singapore: World Scientific Publishing.

Stocking, H. and Holstein, L.W. (1993) 'Constructing and reconstructing scientific ignorance: ignorance claims in science and journalism', *Knowledge: Creation, Diffusion, Utilization*, 15(2): 186–210.

Stocking, H. and Leonard, J.P. (1990) 'The greening of the press', *Columbia Journalism Review*, 29(November/December): 37–44.

Stokstad, E. (2004) 'Salmon survey stokes debate about farmed fish', *Science*, 303 (5655), 9 January: 154–5.

Sturgis, S. (2012) 'North Carolina gets a Texas-sized warning on fracking', *Facing South*, 13 June. Online. Available HTTP: http://www.southernstudies.org/2012/06/north-carolina-gets-a-texas-sized-warning-on-fracking.html (accessed 1 September, 2013).

Sunderlin, W.D. (2003) *Ideology, Social Theory and the Environment*, Lanham, MD: Rowman & Littlefield.

Susskind, L.E. (1994) *Environmental Diplomacy: Negotiating More Effective Global Agreements*, New York and Oxford University Press.

Sutton, P.W. (2004) *Nature, Environment and Society*, Houndmills, Basingstoke: Palgrave Macmillan.

Suzuki, D. (1994a) 'Amazon's "Tarzan" plans epic swim', *Toronto Star*, 4 July: B-1.

Suzuki, D. (1994b) 'Study gives biodiversity a big boost', *Toronto Star*, 5 March: B-6.

Szasz, A. (1994) *Ecopoplism: Toxic Waste and the Movement for Environmental Justice*, Minneapolis, MN: University of Minnesota Press.

Tangley, L. (1988) 'Research priorities for conservation', *BioScience*, 38(7): 444–8.

Tansley, A.G. (1939) 'British ecology during the past quarter century: the plant community and the ecosystem', *Journal of Ecology*, 27(2): 513–30.

Tarrow, S. (1988) 'National politics and collective action: recent theory and research', *Annual Review of Sociology*, 14: 421–40.

Taylor, D.E. (1992) 'Can the environmental movement attract and maintain the support of minorities?', in B. Bryant and P. Mohai (eds) *Race and the Incidence of Environmental Hazards: A Time for Discourse*, Boulder, CO: Westview Press.

Taylor, D.E. (2000) 'The rise of the environmental justice paradigm', *American Behavioral Scientist*, 43(4): 508–80.

'The East is grey' (2013) *The Economist*, 10 August: 18–21.

Thompson, M. (1991) 'Plural rationalities: the rudiments of a practical science of the inchoate', in J. Hansen (ed.) *Environmental Concerns: An Interdisciplinary Exercise*, London: Elsevier Applied Science.

Timasheff, N.S. and Theodorson, G.A. (1976) *Sociological Theory: Its Nature and Growth*, New York: Random House.

Timmer, V. and Juma, C. (2005) 'Taking root: biodiversity conservation and poverty reduction come together in the tropics: lessons learned from the Equator Initiative', *Environment*, 47(4): 24–44.

Tinbergen, J. (1977) *Reshaping the International Order*, New York: Dutton.

Tolba, M.K. and El-Kholy, O.A. (1992) *The World Environment 1972–1992: Two Decades of Challenge*, London: Chapman & Hall.

Tolkien, J.R.R. (2004) *The Lord of the Rings* (50th Anniversary Edition), Boston and New York: Houghton Mifflin Harcourt.

Tolmé, P. (2010) 'The dirty truth behind clean natural gas', *National Wildlife*, 14 May. Online. Available HTTP: http://www.nwf.org/News-and-Magazines/National-Wildlife/Animals/Archives/2010/The-Dirty-Truth-Behind-Clean-Natural-Gas.aspx (accessed 2 September, 2013)

Tuchman, G. (1978) 'Introduction: the symbolic annihilation of women by the mass media', in G. Tuchman, A.K. Daniels and J. Benet (eds) *Hearth and Home*, New York: Oxford University Press.

Udall, J.R. (1991) 'Launching the natural ark', *Sierra* 76(5), September/October: 80–9.

Ungar, S. (1992) 'The rise and (relative) decline of global warming as a social problem', *The Sociological Quarterly*, 33(4): 483–501.

Ungar, S. (1994) 'Apples and oranges: probing the attitude behaviour relationship for the environment', *Canadian Review of Sociology and Anthropology*, 31(3): 288–304.

United Church of Christ Commission for Racial Justice (1987) *Toxic Wastes and Race as in the United States: a National Report on the Racial and Socioeconomic Characteristics of Communities Surrounding Hazardous Waste Sites*, New York.

Urry, John (2010) 'Consuming the planet to excess', *Theory, Culture and Society*, 27(2–3): 191–212.

Urry, J. (2011) *Climate Change and Society*, Cambridge: Polity Press.

Urry, J. (2013) *Societies Beyond Oil: Oil Dregs and Social Futures*, London: Zed Books.

Valiverronen, E. (1999) 'Review of D. Takacs, The Idea of Biodiversity: Philosophies of Paradise [Baltimore: John Hopkins University Press, 1996]', *Science Technology & Human Values*, 24(3): 404–7.

van Koppen, C.S.A. (1998) 'Resource, arcadia lifeworld: nature concepts in environmental sociology', in A. Gijswijt, F. Buttel, P. Dickens, R. Dunlap, A. Mol and G. Spaargaren (eds) *Sociological Theory and the Environment: Proceedings of the Second Woudschoten Conference (Part Two: Cultural and Social Constructivism)*, SISWO, University of Amsterdam.

Veizer, J., Godderis, Y. and François, L.M. (2000) 'Evidence for decoupling of atmospheric CO_2 and global climate during the Phanerozoiceon eon' (letter), *Nature*, 408, 7 December: 698–701.

Verchick, Robert (2013) 'Politics and progress: Will the White House stall its own climate change plans?' *The Hill's Congress Blog*, 25 July. Online. Available HTTP: http://thehill.com/blogs/congress-blog/energy-a-environment/313513-politics-and-progress-will-the-white-house-stall-its-own-climate-change-plans?tmpl=component (accessed 2 September, 2013).

von Weizsacker, C. (1993) 'Competing notions of biodiversity', in W. Sachs (ed.) *Global Ecology: A New Arena of Political Conflict*, London: Zed Books.

Waldie, P. (2013) 'Fractured: UK hamlet becomes unlikely hub for global debate', *Globe & Mail*, 31 July: A1.

Walker, G. (2012) *Environmental Justice: Concepts, Evidence and Politics*, London and New York: Routledge.

Walker, J.L. (1981) 'The diffusion of knowledge, policy communities and agenda setting: the relationship of knowledge and power', in J.E. Tropman, M.J. Dluhy and R.M. Lind (eds) *New Strategic Perspectives on Social Policy*, New York: Pergamon Press.

Walsh, B. (2012) 'Exclusive: how the Sierra Club took millions from the natural gas industry and why they stopped', *TIME Science and Space*, 2 February. Online. Available HTTP: http://science.time.com/2012/02/02/exclusive-how-the-sierra-club-took-millions-from-the-natural-gas-industry-and-why-they-stopped/ (accessed 2 September, 2013).

Warming, E. (with M. Vahl) (1909 [1895]) *Oecology of Plants: An Introduction to the Study of Plant Communities*. Trans. Percy Groom and Isaac Bayley Balfour, Oxford: Clarendon Press.

Waterfield, B. (2013) 'Dutch fury over fracking', *National Post*, 6 July: FP3.

Watt, L. (2013) 'Air pollution takes toll on China's tourism', Yahoo News, August 13. Online. Available HTTP: http://news.yahoo.com/air-pollution-takes-toll-chinas-tourism-070442471.html (accessed 2 September, 2013).

Wayne, A. (2012) 'Fracking moratorium urged as doctors call for health study', *Bloomberg Businessweek: News From Bloomberg*, 18 January. Online. Available HTTP: http://www.businessweek.com/news/2012-01-18/fracking-moratorium-urged-as-doctors-call-for-health-study.html (accessed 3 February, 2014).

Weber, M. (1978) [1922] *Economy and Society: An Outline of Interpretive Sociology* (Guenther Roth and Claus Wittich [eds]), Berkeley, CA: University of California Press.

Weber, M. (1991) *From Max Weber: Essays in Sociology*, London: Routledge.

Weber, M. (1946) [1918] 'Science as a vocation', in H.H. Gerth and C.W. Mills (trans. and ed.), *From Max Weber: Essays in Sociology*, New York: Oxford University Press.

Weitzer, R. (1991) 'Prostitutes' rights in the United States: the failure of a movement', *Sociological Quarterly*, 32(1): 23–41.

West, P.C. (1984) 'Max Weber's human ecology of historical societies', in V. Murvar (ed.) *Theory of Liberty, Legitimacy and Power*, Boston: Routledge and Kegan Paul.

Westell, D. (1994) 'Urban treatment of sewage gets bad marks: study shows 17 cities fail to treat all waste', *The Globe and Mail*, 15 June: A-4.

Wharton, C.R. Jr. (1966) 'Modernizing subsistence agriculture', in M. Wiener (ed.) *Modernization: The Dynamics of Growth*, New York: Basic Books.

'Why not just genetically engineer women for milk?' (2003) *Scoop Politics*, 1 October. Online. Available HTTP: http://www.scoop.co.nz/stories/PO0310/S00003.htm (accessed 2 September, 2013).

Wiener, C.L. (1981) *The Politics of Alcoholism: Building an Arena around a Social Problem*, New Brunswick, NJ: Transaction Books.

Wilber, T. (2012) *Under the Surface: Fracking, Fortunes, and the Fate of the Marcellus Shale*, Ithaca, NY: Cornell University Press.

Wilkins, L. and Patterson, P. (1990) 'Risky business: covering slow-onset hazards as rapidly developing news', *Political Communication and Persuasion*, 7(1): 11–23.

Wilkinson, I. (2001) 'Social theories of risk perception: at once indispensable and insufficient', *Current Sociology*, 49(1): 1–22.

Williams, J. (1998) 'Knowledge, consequences, and experience: the social construction of environmental problems', *Sociological Inquiry*, 68(4): 476–97.

Willis, J. (2013) *U.S. Environmental History: Inviting Doomsday*, Edinburgh: Edinburgh University Press.

Wilson, C.L. and Matthews, W.H. (1970) *Man's Impact on the Global Environment: Report of the Study of Critical Environmental Problems*, Cambridge, MS: MIT Press.

Wilson, E.O. (1986) 'Editor's forward', in E.O Wilson (ed.) *Biodiversity*, Washington, DC: National Academy Press.

Wilson, E.O. (1992) *The Diversity of Life*, Cambridge, MS: Belknap Press of Harvard University Press.

Wilson, E.O. (1994) *Naturalist*, Washington, DC and Cavelo, CA: Island Press.

Witt, W. (1974) 'The environmental reporter on U.S. daily newspapers', *Journalism Quarterly*, 51: 697–704.

Woodgate, G. and Redclift, M. (eds) (2010) *The International Handbook of Environmental Sociology* (2nd edition), London: Edward Elgar.

World Meteorological Organization [WMO] (1985) 'Report of the International Conference on the Assessment of the Role of Carbon Dioxide and Other Greenhouse Gases in Climate Variation and Associated Impacts', Villach, Austria, 9–15 October (Geneva: WMO Report No. 661).

Worster, D. (1977) *Nature's Economy: A History of Ecological Ideas*, Cambridge: Cambridge University Press.

Wright, E.O. (2004) 'Interrogating the treadmill of production: some questions I still want to know about and am not afraid to ask', *Organization & Environment*, 17(3): 317–22.

Wright, O., McCarthy, M. and Bawden, T. (2012) 'David Cameron: Britain must be at the heart of the shale gas revolution', *The Independent*, 11 December. Online. Available HTTP: http://www.independent.co.uk/news/uk/home-news/david-cameron-britain-must-be-at-the-heart-of-shale-gas-revolution-8406432.html (accessed 3 February, 2014).

Wynne, B. (1992) 'Risk and social learning: reification to engagement', in S. Krimsky and D. Golding (eds) *Social Theories of Risk*, Westport, CT: Praeger.

Wynne, B. (2002) 'Risk and environment as legitimatory discourses of technology: reflexivity inside out?' *Current Sociology*, 50(3): 459–77.

Wynne, B. and Mayer, S. (1993) 'How science fails the environment', *New Scientist*, 138(1876): 33–5.

Yearley, S. (1992) *The Green Case: A Sociology of Environmental Issues, Arguments and Politics*, London: Routledge.

Yearley, S. (2002) 'The social construction of environmental problems: a theoretical review and some not-very-Herculean labors', in R.E. Dunlap, F.H. Buttel, P. Dickens and A. Gijswijt (eds) *Sociological Theory and the Environment: Classical Foundations, Contemporary Insights*, Lanham, MD: Rowman & Littlefield.

Yearley, S. (2005) *Cultures of Environmentalism: Empirical Studies in Environmental Sociology*, Palgrave Macmillan.

Yearley, S. (2009) 'Sociology and climate change after Kyoto: what roles for social sciences in understanding climate change?', *Current Sociology*, 57(3): 389–405.

Yearley, S. (2010) 'Science and the environment in the twenty-first century', in M.R. Redclift and G. Woodgate (eds) *The International Handbook of Environmental Sociology* (2nd edition), Cheltenham: Edward Elgar.

York, R. and Rosa, E A. (2012) 'Choking on modernity: a human ecology of air pollution', *Social Problems*, 59(2): 282–300.

York, R., Rosa, E.A. and Dietz, T. (2003) 'A rift in modernity? Assessing the anthropogenic sources of global climate change with the STIRPAT model', *International Journal of Sociology and Social Policy*, 23(10): 31–51.

Zager, E. (2010) 'Nothing shale low about George Mitchell', *Rock the Capital*, 22 September. Online. Available HTTP: http://www.rockthecapital.com/09/22/nothing-shale-low-about-george-mitchell/ (accessed 2 September, 2013).

Name index

Subject index